THE COMPLETE BOOK OF
UNDERGROUND HOUSES
HOW TO BUILD A LOW-COST HOME

· ROB ROY ·

STERLING PUBLISHING CO., INC.
NEW YORK

5361571

Acknowledgments

The best part about writing this book was renewing old acquaintances (and making new) with all the slightly out-of-step eccentrics who dare to do something radically sensible. Thanks to architects Malcolm Wells, Don Metz, and Les Boyer for the helpful chats, and to Ray Sterling at the Underground Space Center, Susan Nelson at the American Underground Space Association, and Peter Carpenter at the British Earth Sheltering Association. Special thanks to owner-builders Peter and Eileen Allen (for help with the radon commentary), Richard and Lisa Guay, Elaine Rielly Cosgrove, Becky Gillette and Roger Danley, and Linda and Ray Hurst for sharing their experiences. Extra-special thanks go to Geoff Huggins, Siegfried Blum, and John and Edith Rylander for actually writing their own excellent case studies.

Lovingly dedicated to Jaki, my building partner at Log End Cottage, Log End Cave, Earthwood, Mushwood, and who knows what in the future. Thanks for the memories past and all those yet to come. It's been fun.

Library of Congress Cataloging-in-Publication Data

Roy, Robert L.
   Complete book of underground houses : how to build a low-cost home/
by Rob Roy.
     p.   cm.
   Includes bibliographical references and index.
   ISBN 0-8069-0728-2
   1. Earth sheltered houses—Design and construction.   I. Title.
   TH4819.E27R678   1994
   690'.837—dc20   94-16840
                                          CIP

10 9 8 7 6 5 4 3 2 1

Published by Sterling Publishing Company, Inc.
387 Park Avenue South, New York, N.Y. 10016
© 1994 by Robert L. Roy
Distributed in Canada by Sterling Publishing
% Canadian Manda Group, One Atlantic Avenue, Suite 105
Toronto, Ontario, Canada M6K 3E7
Distributed in Great Britain and Europe by Cassell PLC
Villiers House, 41/47 Strand, London WC2N 5JE, England
Distributed in Australia by Capricorn Link (Australia) Pty Ltd.
P.O. Box 6651, Baulkham Hills, Business Centre, NSW 2153, Australia
Manufactured in the United States of America

Sterling ISBN 0-8069-0728-2

Contents

Introduction

by Malcolm Wells

Rob Roy is a man you can trust, and I don't say it just because he has returned the books I lent him. His honesty is obvious. Glance at any page in this book if you want proof. He's so open about the few errors he's made during his "underground" career it's almost embarrassing to read about them. But then, as I read on, I find his openness reassuring, his solutions simple and convincing.

Rob is one of a growing number of people who've built far more underground houses than I, yet he assumes that I've done far more than I have. Rob's not only talked about underground architecture, and written about it, he's actually gotten out there and done it—and taught a lot of others to do it, too.

If you have even a slight interest in building a gentle, unobtrusive kind of house, read on. Rob will take all the mystery out of it. He'll even let his clients tell of their adventures with his designs.

Wells is an architect in Brewster, Mass. His latest book is Infra Structures.

4

Preface

Over the past 20 years, my career has consisted of building, writing and teaching about two rather distinct building styles: cordwood-masonry construction and underground housing. Cordwood masonry, wherein walls are constructed of short logs laid transversely like a stack of firewood, will be only briefly touched upon in this volume. However, many of the builders featured in the case-studies section did incorporate cordwood masonry into their homes. Those interested in more information on cordwood masonry should refer to the *Complete Book of Cordwood Masonry Housebuilding: The Earthwood Method* (Sterling Publishing Co., Inc., 1992). That book also gives a thorough step-by-step account of the construction or our own round cordwood-masonry earth-sheltered home, Earthwood.

Underground (also known as "earth-sheltered") housing is a broad subject. In the late 1970s and early 1980s, there was a national mini-craze about underground housing, probably in response to heightened awareness of energy conservation. Several new books on the subject were being published each year.

Some of the best of these books were produced by the Underground Space Center at the University of Minnesota. The Center is still in existence, but it has de-emphasized housebuilding, probably in response to diminished national interest. Its attention has recently been focused on other uses for underground space: storage, industrial and commercial use and tunnelling. In a recent conversation with Ray Sterling, USC's director, who also co-authored many of the Center's books, I learned that interest in earth-sheltering is picking up again.

Malcolm Wells, architect, writer and visionary in the underground-housing movement, continues his work in Brewster, Mass. Wells has been at it since 1964!

My own *Underground Houses: How to Build a Low-Cost Home* (Sterling Publishing Co., Inc., 1979) sold over 100,000 copies in twelve printings, but has been out of print for several years. This present volume contains much material from that work, now greatly revised and updated.

Underground housing is a broad subject, much broader than cordwood masonry, for example. I've found it necessary to specialize within the larger field, and accent methods and techniques that make underground housing both doable by and affordable to the owner-builder who hasn't won the state lottery. The methods accented here are: the slab foundation, surface-bonded block walls, post-and-beam and plank-and-beam framing, and the self-sealing rolled-out waterproofing membrane. These are the methods that my wife, Jaki, and I have used successfully at Log End Cave, Earthwood, and several earth-covered outbuildings.

This book uses Log End Cave as a model, and it takes the reader through the step-by-step construction of a single-storey rectilinear low-cost underground home. Where improvements to the design are suggested, attention is called to the 40' × 40' Log End Cave plans on page 15. The case-studies section expands on the basic theme somewhat by including houses of other shapes, and shows examples from the "Deep South" of the U.S. to Ontario, Canada.

The reader, if he intends to build his own house, should design it himself (or herself) too. This is one of only two ways to get the house you want. The other way is to hire a good architect.

There's no danger of my putting contractors or architects out of business by advising that folks design and build their own homes. While the methods and dimensions described in this book have worked well for us, it's important to have one's plans checked by a competent architect or structural engineer. After all, conditions of soil, climate and materials vary from site to site. But having plans checked or critiqued is much less expensive than handing an architect a blank sheet and saying, "Design a house that will be right for me." Sometimes it's possible to have stress-load calculations checked by a college's structural-engineering class as a study project.

As for construction, even the most resourceful owner-builders may feel that there are building trades that are simply beyond their confidence or ability. Even here, though, it's possible to save money by acting as one's own contractor, sending work write-ups out to subcontractors for bidding,

and getting discounts at the local building-supply yards. Some readers may subcontract almost all of the work, and still derive substantial savings over a contractor-built home.

Finally, our own Earthwood Building School has since 1980 acted as a clearinghouse for information about earth-sheltered housing and cordwood masonry.

—Rob Roy
West Chazy, N.Y.

1
Past & Present

Underground housing, even *low-cost* underground housing, is nothing new. After all, what did it cost the first caveman to walk into his new shelter? He certainly didn't have a thirty-year mortgage.

Although there's some very posh housing converted from true caves in the Loire Valley in France, modern earth shelters have little in common with caves.

Man has lived underground at many times and places throughout history. In Cappadocia, Turkey, people have been living in underground towns and cities for thousands of years. The settlements, some of which extend eight stories below ground, are hewn out of soft stone. These settlements are a response to a hostile surface environment, and the microclimate of the underground villages remains constant and comfortable, despite harsh variations of heat and cold on the ground's surface.

Folk architecture has always made intelligent use of available materials. When these consist of no more than the soft strata itself, it's only natural that underground dwellings evolve. In China, this type of development has persisted for 6000 years. In *Earth Sheltered Housing Design*, we learn:

In 4000 B.C. villagers at the recently unearthed Banpo site in China lived in semiunderground pit dwellings with an A-frame roof supporting a thin layer of soil and vegetation. In later centuries cave dwelling became a widely adopted practice in the arid regions of eastern central China, where a deep loess soil provided ideal conditions for self-supporting excavations in soil. Marco Polo . . . noted several tribes who lived in excavated homes. The thermal advantages of the earth's temperature moderation in a continental climate with cold winters and hot summers, together with the ability to provide shelter with only a pick and shovel, have led to the construction of millions of these

cave dwellings through to the present day. In many locations, farming continues on the surface above the below-grade houses. Approximately 20 million people live in cave dwellings in China today.[1]

On Tenerife, in the Canary Islands, cave-dwelling is found in the agricultural uplands. When I visited this area, I thought these homes to be most uncavelike. It was as if houses had been set into the hillside, with only one wall left exposed. There were even curtains on the windows. The houses weren't brightly lit (as it is possible to make them today with the advent of the insulated skylight) but they were cheery and will no doubt be much the same hundreds of years from now. How many generations of little troglodytes returned to their cool homes after a hot day on the terraces?

Why would anyone want to build underground in the modern world? Malcolm Wells says:

Every square foot of this planet's surface—land and sea—is supposed to be robustly alive. It is not supposed to be shopping-centered, parking-lotted, asphalted, concreted, condo'd, housed, mowed, polluted, poisoned, trampled, or in any way strangled in order that we—just one of a million species—can keep on making the same mistakes.[2]

Every house which we can design and build has an impact of some kind on the planet, usually negative. The type of home with the potential for the least negative impact is the underground home. It's the only kind of housing that allows a return to something approximating the original landscape.

A great example of this housing is a beautiful home in Yorkshire called "Underhill." It was designed and built by Arthur Quarmby, architect and plastics engineer. Having lived in Britain for seven years, I knew that one of the primary concerns of the county and regional planning boards was to retain the "visual amenity" of the villages and countryside. A much better job of retaining this has been done in the U.K. than has been done in the U.S. Quarmby's house, at that time, was the only fully earth-sheltered home in England (a handful have been built since 1980), largely because of resistance by the planners! This is very difficult to understand. At "Underhill," the same sheep which had been grazing the Quarmby lot on the edge of the village were able to come back to continue their grazing after the home was finished. From the main road, the home is hardly seen. That's visual amenity . . . not to mention agricultural and environmental integrity (Illus. 1–1).

Jaki and I once lived in an old stone cottage on the outskirts of Dingwall in Scotland. We were in the middle of a 600-acre sheep-and-cattle farm on the south-facing slope of a long hill. Over the years, the town of Dingwall gradually made its way towards the farm where we lived, one field at a time falling to tract housing. The tall (almost A-frame) houses that were built were crowded 10 to the acre in the development, the net effect being that the "view" from any house was the gable ends of the houses facing it. Had a development of "Underhills" been approved instead, the view from each home would have been of the grassy roof of the home facing it, leaving an unobstructed view across the Cromarty Firth. The sheep could have returned.

Architerra, an international architectural group specializing in earth-sheltered housing, actually built several complexes in France and Spain that would have been perfect in place of the imposing homes that were built near Dingwall.

By virtue of the stepped design of the Architerra complexes, and the curved-glass front walls of the units, all of the residents enjoy a 180° view of the surrounding landscape. These south-facing glass walls provide natural light and view and passive solar gain and help the units appear naturally integrated with the hillside. Stacking the units stair-fashion in slots cut into the hillside imparts a sense of privacy to each unit—because of the unobstructed view of the horizon—while permitting a high unit density.[3]

In Architerra's Nice project, the density was about 10 houses per acre, similar in density to the aforementioned above-ground project in Scotland. Architerra tried to launch similar projects in the United States in the 1980s, but their great ideas went unfunded.

Many readers who've followed (or tried to follow) the underground movement since the heady days of the late 1970s may wonder what has happened in the past ten or fifteen years. Malcolm Wells, as stated, is still active, even after three decades in the movement. Ray Sterling and John Carmody, authors of *Earth Sheltered Housing Design*, still work at the Underground Space Center at the University of Minnesota, although their work has more to do now with underground space for uses other than housing. Architect Don Metz, who designed and supervised the construction of some of the most beautiful earth-sheltered homes I've seen, is still active, as is Les Boyer, Professor of Architecture at Texas A&M University. Andy Davis, who publicized his "Davis Caves" so well in the 1970s is still going strong. The British Earth Sheltering Association is probably more active in promoting residential uses of underground housing than any comparable organization in America, although only a handful of earth shelters have actually been built in Britain.

The bad news: The magazine *Earth Shelter Digest* was a casualty of the 1980s, along with Architerra and several other construction-and-design companies which tried to make a living at specializing in earth-sheltered housing.

The good news: There's a sense of optimism among the old guard that underground is on the way back. Interest has picked up in Malcolm Wells' books, and, after a period of slow going, at the Underground Space Center as well. Interest is increasing again at Earthwood Building School, where we've conducted underground-housing workshops continuously since 1980.

What caused the bust which followed the boom? Why is interest returning?

Publicity in the 1970s followed a new awareness that energy sources weren't inexhaustible. The public took a fancy to the "new" idea of underground houses. Articles in magazines such as *New Shelter* and *The Mother Earth News* proliferated. Everyone seemed to be publishing a book on underground housing. I've got dozens of different volumes on my reference shelf.

Then came the public perception that the energy crisis was either over, or had been a fraud in the first place; neither view turned out to be true. President Reagan appealed to upward mobility and unbridled economic optimism. While the rich got richer, the poor got poorer. The middle class is only now realizing that the same old realities are still with us, to wit: We still waste inordinate amounts of energy in the United States when compared with other industrial nations with a similar standard of living. We continue to encroach further and further upon the very ecological systems that support life on this planet: open water and wetlands, air, forests, topsoil.

For the first time in a decade, young people seem to be concerned again with the environment, with quality of life instead of standard of living. Underground housing is very much in tune with this thinking.

Illus. 1–1. Arthur Quarmby's "Underhill" home, in Holme, Yorkshire, U.K., maintains the visual amenity of the village and countryside.

2
Design

BERMED VERSUS TRUE UNDERGROUND

There are two different approaches to earth-sheltered housing: the bermed house and the chambered (or truly "underground") house. The bermed house involves building the structure at or close to original grade and "berming" (mounding earth against) the side walls. Very often, an earth roof is chosen to complete the harmony of the building. In the chambered house, the entire structure is below original grade.

There aren't many "true" underground homes in the United States, although there's a great deal of underground commercial space. The only such home I've visited is architect John Barnard's first Ecology House in Marstons Mills, Mass. A below-grade central courtyard provides access to several underground rooms opening onto it. The courtyard, in turn, is accessed by a single stairway down from ground level. The approach is very similar to that taken in China, where individual homes are carved out of the loess subsoil, all accessed to a central courtyard.

In later designs, John Barnard's courtyard evolved into a covered atrium. The below-grade aspect of the original home was tempered somewhat by more of a bermed approach, with a south-facing elevation providing access and light onto grade.

In addition to the courtyard and the covered atrium, leaving one or more sides of the home exposed to grade (this is called the "elevational" approach to earth-sheltered housing) is another way to provide ingress to the home, natural light, and ventilation. Yet another way is to use sidewall penetrations through the earth berm as door and window locations. An example of this is the fanciful "hobbit" door (Illus. 2–1) in Arthur Quarmby's "Underhill" home in Yorkshire, in the United Kingdom.

Illus. 2–1. A penetrational entranceway

Many different techniques have evolved over the past 30 years which make underground houses as light and bright and airy as those homes built above ground. Underground designers seem to go out of their way in this respect, cognizant not only of the code-enforcement officer's strict ad-

10

herence to building regulations, but also of the common citizen's belief that underground housing equals dark, damp, dingy basements. Mike Oehler says:

An underground house has no more in common with a basement than a penthouse apartment has in common with a hot, dark, dusty attic.[4]

I remember my visit to "Baldtop Dugout," architect Don Metz's earth-sheltered home. The 270° panoramic view of the surrounding New Hampshire mountains and Connecticut River Valley dispelled once and for all the notion that underground houses are lacking in views. Our own Log End Cave had a wonderful close view into the woods, where the activities of the local wildlife seemed to be almost a part of the living space.

THE LOG END CAVE DESIGN

At Log End Cave, we decided to compromise between the bermed and chambered styles. We would use material excavated for the foundation to build up the east and west berms for the home.

We would also use a semi-bermed south elevation to provide access, lots of light, and a view. I must confess that there are three glaring design faults on this original Log End Cave design's south elevation. They are:

- No thermal gain is accomplished by berming up to the underside of the windows on the south side. It just looks nice. The drawback is that snow starts to accumulate right there, and, in northern New York, it isn't long before somebody is out there with a snow shovel. Also, every square foot of a south-facing elevation given over to double-pane insulated glass will actual provide a net energy gain in northern climes.
- We suffered from an energy "nosebleed" where the east- and west-side block walls conducted the home's internal heat directly to the outside. We corrected the problem with some retrofitted polystyrene insulation on the exterior, but this should have been attended to at the design stage.
- The south elevational entrance is the only means of ingress and egress. Even a chipmunk knows better. One never knows when a fox (or a building inspector) might come to the door. Have a

second means of escape. Now, this isn't purely a design fault of the south-elevational wall alone. There are other ways to incorporate the code-mandated (and sensible) second entrance besides placing a second door on the elevational wall: a penetrational doorway through the berm, for example.

But, the elevational plan did fit in well with the site, despite the errors cited above. People have commented that the Cave seemed to fit the natural terrain better than most other earth-sheltered homes. We felt that the atrium and sidewall penetration design techniques weren't as well suited to providing the external view that was so important. These techniques would also involve expenses that we weren't prepared to make at the time.

The Log End Cave which we actually built in 1977 was about 30' × 35'. I include these plans (Illustrations 2–2, 2–3, and 2–4) as well, although they are inferior, to illustrate certain points in the narrative. The dimensions were a function of site considerations, affordability, availability of materials, and certain goals for heating and cooling efficiency. The floor plan featured an open living/kitchen/dining area, and smaller rooms on the east and west sides, very much like the 40' × 40' plans shown on page 15. For structural safety with 4 × 8 rafters, we limited ceiling spans to about 8'6". This made the perimeter rooms rather small, a shortcoming which was corrected in the 40' × 40' plans, based on ten-foot-square sections. When we visit Richard and Lisa Guay's home, based on those plans and described in chapter 16, we're impressed by the comparatively roomy bedrooms and bathroom with their ten-foot spans.

Although every owner-builder likes to design his own home, incorporating the features which he finds important, I offer a 40' × 40' Log End Cave plan (Illustrations 2–5, 2–6, and 2–7) to show certain construction techniques common to underground homes. The house can be built on a gentle slope, as ours was, or on a flat site as a bermed structure.

At the original Cave, our north-south dimension (30') was limited by the availability of three 30' 10" × 10" barn beams. At the time, we didn't realize that overall structural strength isn't compromised by joining shorter girders over the pillars. With ordinary light-frame construction, bending

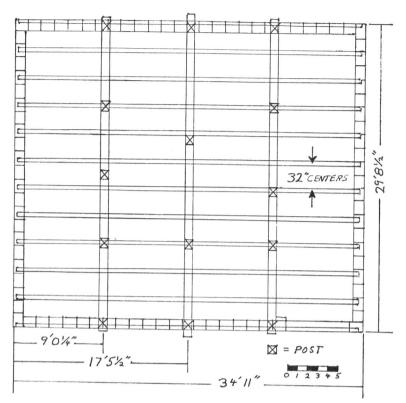

Illus. 2–2. The original Log End Cave's block, rafter, post-and-beam plan

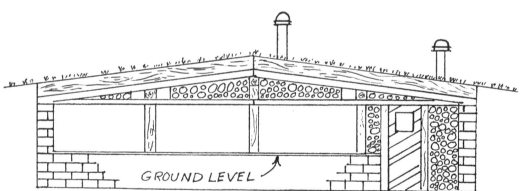

Illus. 2–3. The south-wall elevational plan of the original Log End Cave, a bermed structure built on a gentle slope

strength generally comes into play before shear strength, but on heavy-timber construction used for heavy loads, the opposite is usually the case. (*Bending failure* is when a member snaps somewhere near the middle because of the load. *Shear failure* is the tendency of all the fibres of the wood to "shear" through, usually right near where the member is supported by a wall or a post.) Two separate 10' girders joined over a post are actually stronger on shear than a single 20' girder supported halfway along by a post. While bending strength is slightly decreased in this example, shear strength increases by about 20%, and shear strength is usually the weak link in the calculations. Be happy that you only have to deal with 10' girders, not 20' or 30' behemoths.

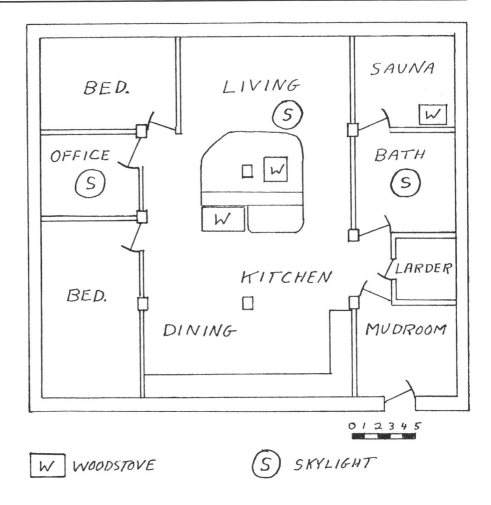

Illus. 2–4. The original Log End Cave's floor plan

BED.

LIVING (S)

SAUNA (W)

OFFICE (S)

BATH (S)

W

W

BED.

KITCHEN

LARDER

DINING

MUDROOM

0 1 2 3 4 5

W WOODSTOVE   S SKYLIGHT

The 40′ × 40′ floor plan (Illus. 2–7) will result in peripheral rooms that are a few degrees cooler than the open-plan great room, assuming that you have centrally located wood heat. This was true at the original Cave, and was considered at the design stage. Jaki is English, and I had lived seven years in Britain, so we were used to cool bedrooms, and believe that they are healthier than their overheated American counterparts. Peripheral-room temperatures can be further regulated by opening or closing the internal doors. For nonwoodburners, perimeter baseboard heating with zoned thermostatic control is always an option. Finally, I always design floor plans to make the joining of internal walls with exposed rafters and girders both neat and easy to construct.

Our east-west dimension of about 35′ was limited by the strength of 4″ × 8″ rafters. The 40′ × 40′ plans call for 5 × 10s, as we used at our round Earthwood house. In 10′ to 12′ lengths, they're heavy, but not unmanageable. You'll want some help with the heavy timbers. The stress-load engineering for the plans is based upon a 10′ × 10′ module, repeated 16 times. Therefore, the plans are easily adaptable to other sizes and shapes: 30′ × 60′, 20′ × 40′, etc.

HOW AN UNDERGROUND HOUSE WORKS

Many people casually acquainted with underground housing think we build this way to take

*Illus. 2–5. The new (40' × 40')
Cave's block, rafter, post-and-
beam plan*

*Illus. 2–6. Below: The 40' × 40'
Log End Cave plan improves upon
the original design. The home is
designed to support a roof load of
at least 150 lbs. per square foot.*

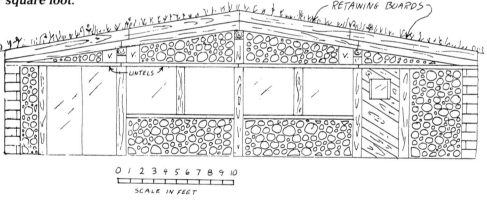

advantage of some great insulative value of earth. *This isn't true!* In fact, earth is a pretty poor insulator, and wet earth is a *terrible* insulator.

So how do underground houses save so much energy in heating or cooling if earth is poor insulation? The earth is a great *capacitor.* Just as an electrical capacitor stores an electric charge, the earth is a capacitor which stores heat. For us, building near Plattsburgh, N.Y., building the house 6' to 8' below grade would be like building it 1000 miles to the south, with a winter climate more like that of Charleston, S.C. The ambient earth tem-

perature just outside the house walls in winter is about 40°F (Illus. 2–8). When the outside winter air temperature is −20°F, as is often the case, the underground house starts out with a 60°F advantage over the house on the surface. Put another way, the underground house need only be 30°F warmer than the ambient temperature to reach a comfort level of 70°F. Meanwhile, the surface dwelling needs to be 90°F warmer than its ambient, the frigid outdoor air.

While the earth is a good capacitor, it's also a good conductor. The 40°F earth will try to rob the

Illus. 2–7. This 40' × 40' Log End Cave floor plan can be altered to meet individual needs. The key is to conform to the 10' × 10' square modules for which the home is engineered.

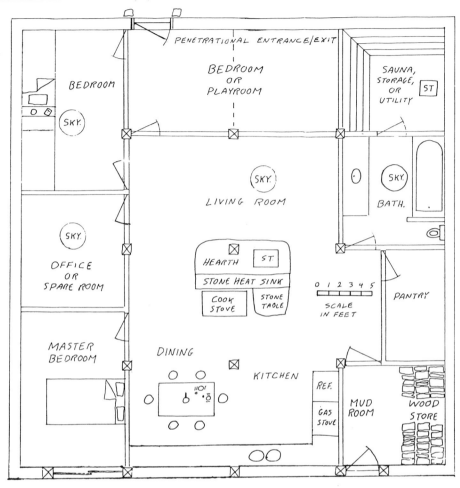

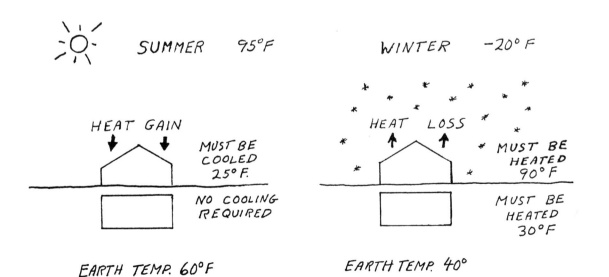

Illus. 2–8. The heating and cooling advantages of an underground house

heat from the house, and an internal temperature of 40°F won't be very comfortable. Thankfully, there's another thermal mass at our disposal, one over which we can more easily exercise control. This is the mass of the fabric of the building itself: the concrete floor, the walls, the footings, and any internal mass such as a masonry stove. The best way to regulate this thermal mass is to separate it from the earth's thermal flywheel by a *thermal break,* typically rigid-foam insulation placed correctly on the exterior of the home's fabric. Now the house itself can be brought up to temperature and the advantage of the earth's favorable ambient climate can be utilized. This is very important. The worst thing to do (and it's amazing how often it's done) is to place insulation on the interior of the thermal mass. Not only has all control over the mass (perhaps 100 tons) of the home's fabric been lost, but now the earth can freeze up against the cold walls of the home and cause structural damage.

In the summer, there's a similar advantage. Think of the earth as storing "coolth." In northern New York, the highest earth temperatures are about 60°F at six feet deep. This peak occurs in August, after months of slow, steady rise after the 40°F earth temperature in early March. Even if it's 95°F outside, no energy is required for cooling. Residual heat in the home, sunlight, people heat (98.6°F), dogs, lighting, and cooking will all bring the house up to maybe 75°F. The house built on the surface has two choices: stifling heat or energy-draining air conditioning.

If an earth roof is included in the design, there's another great cooling advantage. Unlike the high surface temperatures of asphalt tarscapes, the earth roof is cool just a few inches below the surface. Respiration by plants and evaporation of moisture off the earth roof both help to cool the building, just as a wet towel draped over a 5-gallon bucket will help to keep your drinks cold.

EGRESS CODE & FIRE SAFETY

The 40′ × 40′ Log End Cave plan is an example of elevational terratecture with bermed sidewalls.

Although both external doors are on the elevational side, there's no reason why a penetrational door couldn't be incorporated on one of the other walls if access were required in a different direction, or if you encountered building-code difficulties. Penetrational bedroom windows, if they open, can also satisfy building codes egress. For example, the National Building Code (NBC) allows the egress windowsill to be no more than 48″ off the floor. Other codes may specify 44″. Typically, rescue windows from sleeping rooms must have a minimum net clear opening of 5.7 sq. ft. The minimum net clear-opening height of 24″ and the net clear-opening width must be at least 20″. Some building codes will allow the alternative of two doors from bedrooms offering two separate paths of escape. Ray Sterling and John Carmody say, in *Earth Sheltered Housing Design*:

The intent of the egress requirement is clear. If a fire should start in any part of the house other than the bedrooms, occupants should have a clear means of escape directly to the outside without going into a smoke- or fire-filled part of the house.[5]

While earth-sheltered houses are very much less likely to catch fire due to their use of massive materials such as concrete, concrete blocks, and heavy timbers, it's still possible for dangerous smoke fires to occur in furniture, or electrical fires in internal walls. Meeting the requirement of building codes may sometimes involve overbuilding and greater expense, but it can also eliminate a lot of unpleasant hassles. The bedrooms in the 40′ × 40′ plans need either interconnected doorways or penetrational code-worthy windows to pass an inspection. Check with your local code-enforcement officer. Or, if you're building in one of those few remaining rural areas where building codes are rare, use your own best-informed judgment. Err on the side of safety.

When designing an underground house, keep in mind three important words: *strength* (because of heavy earth loads on the roof and sidewalls), *livability,* and *waterproof.* You should have enough light and ventilation to assure a pleasant, open, nonclaustrophobic atmosphere. As for keeping the house dry, drainage is the better part of waterproofing.

3
Siting & Excavation

SOLAR ORIENTATION

If you haven't bought your property yet, there are a few considerations special to underground houses which might be less important with on-grade homes.

In the northern U.S. and Canada, the ideal site would feature a gentle south-facing slope that takes advantage of solar gain. Stu Campbell in *The Underground House Book* says:

The perfect exposure for a window meant to collect solar radiation is 15° west of true south, but 20° to either side of this point is still excellent.[6]

In the southern U.S., where cooling the home may be the more important energy consideration, a north-facing slope might be preferable. Remember, too, that north light is much less harsh than south light. East and west-facing sites wouldn't be so bad in the midsection of the U.S., but decide if you're a morning or evening person, as the quantity of light will vary a great deal in different parts of the house from sunrise until sunset. The use of the rooms come into play here when you design the floor plan. As the sun begins to set, a great deal of heat comes in from west-facing windows.

SLOPES

Gentle slopes are good for single-storey homes, which would include the majority of earth shelters that have been built. For a two-storey home, a steeper slope would work, but much more atten-tion must be paid to lateral stress on the walls. Positive drainage is particularly important, especially if the site is on the side of a long hill. In this case, a surface drainage ditch with perforated drain tile near the bottom and filled with 3″- to 4″-diameter stone is installed on the uphill side of the home to carry runoff away from it. Typically, these surface drains are 12″ to 18″ wide and 36″ to 48″ deep, depending on frost depth and other considerations peculiar to the site. On any steep-slope projects, consult with a soil- or structural engineer. Flat sites can also work, provided they have good drainage characteristics.

The Log End Cave homestead featured a knoll higher than any point within a quarter-mile. This knoll sloped gently in all directions, but, luckily, its greatest slope was angled almost due south. The site combined the advantages of good drainage, away from the building, with southern exposure for solar gain in the winter. An added benefit was that the site was closer to our windplant than Log End Cottage had been. With a 12-volt energy system, short cables are best, as there's considerable line loss at low voltage. Finally, driveway access was excellent. Building materials and ready-mix concrete were easy to deliver to the site.

The site had been a meadow when our homestead had been part of a hilltop farm many years ago, but this particular corner of the meadow was overgrown with small apple, cherry, and poplar trees. It took Jaki and me a couple of days to clear enough of the growth to see the terrain clearly. But we'd need a more accurate understanding of the contours than we could discern visually. Front-end loaders cost $22/hour back then, a great deal of money for us in 1977. Plan ahead so that just the right amount of earth is moved, and not moved twice.

FIT THE HOME TO THE SITE

Our method of siting the house worked well for us, and I'd recommend it to anyone faced with a hilltop or hillside. First, we set up a surveyor's transit at the top of the knoll. You can rent a contractor's level (which will do the same job) quite inexpensively at tool-rental stores. We plumbed and levelled the transit to a point on the ground within the legs of the tripod and marked the spot with a half-brick. An even better benchmark would be a large nail driven into the ground, with a red piece of tape tied just under the head so the nail can be found again.

Establish true south and drive in a stake along the true-south alignment. South would then correspond to 180° on the compass rose of the transit.

The easiest way I know to find south is to read the paper or listen to the evening weather, and learn the times of sunrise and sunset on the day you're working. Halfway between those times the sun will be at true south. Cloudy for a week? Use a magnetic compass, but don't forget to take into account the magnetic deflection for your area. A good transit will have a magnetic compass built in, but a rented contractor's level probably won't.

At Log End Cave, we set the zero-degree mark of the level's compass rose to a fixed point, the corner of Log End Cottage. Then, with Jaki holding a calibrated grade stick (you can rent or make one) and carrying the end of a 50-foot tape, and with me reading the level and marking the distance on a chart, we statistically mapped the area.

The procedure was to establish a slope for each

Table 1
Statistical Abstract of the Building Site

*Contour	−90°	−75°	−60°	−45°	−30°	−15°	0°
1″	23′7″**	16′5″	12′2″	9′4″	7′7″	6′8″	6′1″
6″	33′4″	29′4″	24′9″	21′6″	18′6″	15′11″	15′
12″	36′	34′2″	32′8″	27′11″	22′7″	21′	19′4″
18″			36′	31′10″	27′11″	26′	24′2″
24″				36′1″	33′	30′2″	27′10″
30″				41′4″	37′	32′4″	30′4″
36″				46′4″	40′10″	38′2″	35′2″
42″					47′	42′5″	40′6″
48″					54′	49′2″	46′7″

	+15°	+30°	+45°	+60°	+75°	+90°
1″	6′	6′3″	6′6″	7′	7′8″	8′
6″	15′7″	16′2″	17′2″	16′7″	17′	21′6″
12″	22′6″	21′7″	20′11″	21′10″	27′2″	29′
18″	24′4″	23′	22′5″	26′3″	31′4″	35′8″
24″	26′6″	25′10″	25′11″	30′1″	36′	40′8″
30″	29′9″	29′1″	29′4″	34′	38′10″	45′3″
36″	34′8″	32′8″	33′7″	37′5″	42′4″	49′7″
42″	38′2″	36′8″	36′3″	39′5″	46.′4	
48″	43′1″	40′10″	40′1″	41′9″		

*Contours in inches below reference point.

**Distance from reference point to certain contours along primary rays.

ray of the circle divisible by 15°: 0°, 15°, 30°, 45°, and so on. I'd set the level or transit at 0°, for example, and Jaki would move away from me with the grade stick until we were able to discern a drop in the land. The stick's calibrations began reading upwards in inches from a point on the stick equal to the height of the level off the ground, say four feet. As soon as we could perceive an inch of drop, we took a measurement from the benchmark and recorded the distance. Then Jaki moved further away

on the same ray until we read a 6″ drop. Again, we measured the distance. Likewise, we recorded every 6″ drop along the ray until we were outside the vicinity of the house site. Then we repeated the procedure along the 15° ray, and so on to 180°. We now had a statistical abstract (Table 1) of the large general area of the building site. This job took us all afternoon.

I took the transit inside for the evening and transcribed the figures onto a large piece of graph

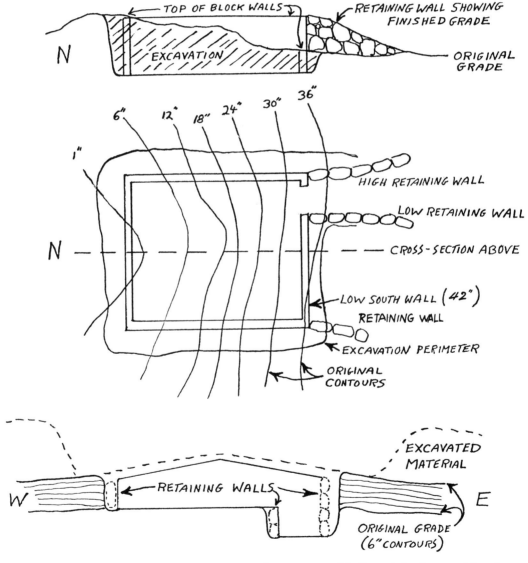

Illus. 3–1. The north and south walls were almost even with the original ground level; the terrain around the east and west walls was built up using excavated material.

paper. I let ⅕″ (the size of the squares on the graph paper) equal one foot. I drew a light pencil line at every 15° of arc for the half-circle in which the house would surely fall, and measured the scaled distances from the center point, marking with a dot each 6″ drop in contour along the way. To avoid confusion later, I lightly marked the elevation next to the corresponding dot: −1″, −6″, −12″, and so on.

We considered −1″ to be level for our purposes. After all the figures from my chart were thus transcribed, it was an easy task to connect the dots of like elevation with gently curving lines. Lo and behold, an extremely accurate contour map of the site emerged as the dots were connected (Illus. 3–1).

All this may sound like a lot of work, but it paid off by eliminating a lot of guessing later on. From a piece of the same kind of graph paper, I cut out a scale model of the foundation plan, using the outside dimensions of the planned block wall. I noted the location of the door on the south wall of the scaled drawing. Now all we had to do was slide the little square around the contour map until the most sensible location emerged.

We knew that the top of the block wall on the south, where there would be large windows, would be 42″ below the top of the other three walls. We knew that the concrete slab at door level would be 78″ below the tops of the three full-height walls. As seen in Illus. 3–1, the north and south walls are about even with the original grade. The south wall required 6″ additional excavation. The earth that comes out of the hole during excavation is used to build up the terrain to the tops of the east and west walls. By this plan, the new ground level melds nicely into the shallow-pitched, earth-covered roof.

Our job, then, was to plot the location of the excavation and to formulate the most efficient excavation depth to take advantage of natural contours, keeping in mind that we'd have to do something with every cubic foot of material that came out of the hole. Drawing a contour map made this very much easier to estimate. It was worth the effort: The landscaping around the Cave wasn't a big job and it came out very well.

MARKING THE EXCAVATION

There's another big time-saver that comes out of these little maps, now that the trouble has been taken to do them: the remarkable ease of placing the four flags to guide the excavation contractor. Once we determined on paper the best location for the house, all we had to do was transpose the four corner points to the site itself. We used a simple angle-and-distance system. Using a small protractor and a ruler, it was easy to determine, for example, that the northeast corner of the house should be 18′3″ from the half-brick used as a benchmark, at an angle of 29°. It's important to be able to set up the transit or contractor's level in exactly the same spot as it was when the original figures were taken. We set up over the brick using the corner of the cottage as 0°, as before, and, at an angle of 29°, we measured out 18′3″ and drove a white-birch stake into the ground (Illus. 3–2). We did the same with the other three corners. Then we checked the wall lengths and diagonals with our 50′ tape, holding the tape level to account for the drop in terrain. Every dimension was within 5″, and after ten minutes of juggling, we put out all the stakes. Walls were the right length; diagonals checked.

I happened to have a 4′ × 4′ sheet of plywood lying around, which I used to establish the corners of the excavation itself (Illus. 3–3). We drove in another set of hefty birch stakes to mark these points. Of course, the original four stakes marking the house corners would be eaten by the front-end loader, but I wanted the operator to be able to visualize the project clearly before starting. It's important that the operator sees the site as the owner-builder does. With our contour map, it was easy to explain just how deep to go at each corner of the excavation.

A 4′ work space all around the walls may seem like a lot, but it's important to have plenty of room to build and waterproof the walls. A certain amount of erosion can be expected during the job, and although we spaced our stakes 4′ out from the house corners in each direction, we didn't end up

Illus. 3–2. Transposing the four corner points from paper to the site, using a protractor, ruler, and surveyor's level

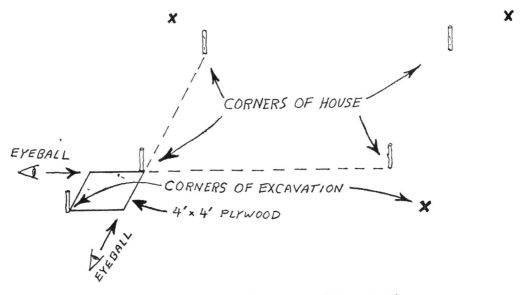

Illus. 3–3. Establishing the corners of the excavation

with anything like 4' of space outside the walls. The space was about 30" on average, at the base of the excavation. If the digger leaves the markers standing in the ground, the excavation slope (called the "angle of repose") will cut away into the four feet at the bottom (Illus. 3–4).

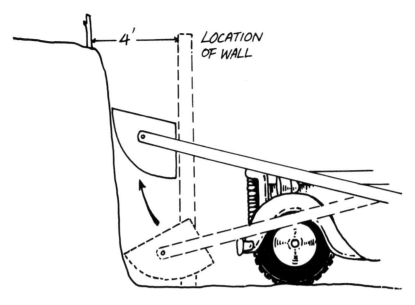

Illus. 3–4. A 4' work space all around the excavation is sufficient room for building and waterproofing the walls.

Illus. 3–5. A shallow angle of repose

If you're fortunate enough to have very sandy or gravelly subsoil (which means that you have excellent drainage), then the angle of repose will probably be more like that shown in Illus. 3–5. Our soil had a fair amount of clay, which allows steep excavation sidewalls, like those shown in Illus. 3–4. With sandy soils, it's necessary to make a bigger excavation, so the second set of stakes might have to be 6' or more out from the house corner stakes. On the positive side, you'll probably be able to backfill with the same material which came out of the hole, because drainage is excellent in sandy soils.

If in doubt about the quality of the subsoil, check with a percolation test and/or a deep-hole test. Your local cooperative extension office or county health department may be able to help with soil maps and information about septic systems. The same tests used to determine percolation for the design of septic systems will give you important information on the drainage characteristics of local subsoils. Since you'll probably have to conduct these tests anyway to satisfy the health department for approval of a septic-system design, you might as well do the tests before you do the site plan. A deep-hole test (5'- or 6'-deep hole) may reveal ledge or bedrock which might affect the whole site plan, or you might find that that wonderful sandy soil is only three feet deep, with hardpan or clay below.

RADON

Radon is a clear, tasteless, odorless gas, which, in sufficient quantity, can cause lung cancer. It enters the home through cracks in the foundation. Underground houses can pose a higher risk than do other types of housing by the nature of their construction. If you're in an area known for high radon concentrations, or if you're building in gravel over shale, granite, or phosphate deposits, or if you simply want peace of mind on this matter, now's the time to test the site for radon. Appendix 1 treats radon in detail, and gives sources of additional information.

AN IMPORTANT QUESTION

The drainage characteristics of the subsoil where you build must be considered carefully when calculating the size of the excavation. You need to answer one very important question: *Can the excavated material be used to backfill the walls of the home?* If percolation in the subsoil is good, then the answer is *yes*. If the soil holds water or doesn't let it through, as with claylike soils, then the answer is *no*. If percolation is poor, you'll have to bring in backfilling material or use one of the various drainage products made for that purpose. These products will be discussed in chapter 10, but the builder should know about them early in the design stage, as the dimensions of the excavation will depend on whether or not it will be necessary to bring in backfilling material. This will probably be an economic decision, weighing the cost of many loads of coarse sand or gravel versus the cost of manufactured drainage materials which are designed to be laid up against the sidewalls of the home.

If you've got poor drainage, read chapter 10 now. If you've got horrendous drainage, such as soils designated as "expansive clays," or if a deep-hole test breaks through to the water table, reconsider the site altogether, or build an above-ground structure by methods which have proven to be successful in your area. Underground housing isn't for everyone's building site.

THE FLAT SITE

If your site is flat, you'll be spared the tedium of creating the contour map. All you need to know is the soil conditions, and I would definitely advise a deep-hole test. The hole doesn't take long for a backhoe to dig, and, you may be required to do one anyway, as is the case in New York State.

For flat terrain, I recommend the bermed style. The Log End Cave design, in fact, is midway between the bermed and chambered styles of underground home. With the bermed style, the builder need only calculate the amount of earth needed to mound up gracefully against the sidewalls. But keep in mind that unless the site is entirely of sandy soil or good-draining gravel, it isn't wise to backfill with the excavated material. Clay and other soils with poor percolation qualities should be kept away from foundations and earth-sheltered walls. Bring in sand or gravel, if necessary, to ensure good drainage. Because our subsoils have such poor drainage, we had to backfill with 25 dump-truck loads of 5 cubic yards each. If it's necessary to bring in backfilling material, this task should be figured in when calculating the depth of the excavation. With a bermed house, it's easy to get rid of a little extra material. The thicker the berm, the better. In fact, if the earth piled up against the walls is thick enough (5' to 6'), the heating advantages are almost the same as they'd be for a house just below original grade.

The two-storey earth-sheltered home where we live now, called Earthwood (see the photos in the color section), was built on a perfectly flat site, a gravel pit. There was no topsoil. We built a pad of good percolating sand, which was right on site, and "floated" the foundation slab on this pad. Forty percent of the cylindrical home was sheltered with an earth berm constructed of gravel pushed up from the area in front of the home.

FLAT-SITE CALCULATIONS

Let's calculate the required depth of excavation for the gabled berm-style house shown in Illus. 3–6. We'll assume poor soil percolation, making it necessary to bring in backfilling material. If the site has good topsoil, the whole area should be scraped by a bulldozer and the soil piled where it will be out of the way. This task will save time and money later when the roof and final landscaping are done. This hypothetical site isn't blessed with a good depth of topsoil. Allow for additional topsoil if you need it at your site. How much material of the kind taken from the excavation will be required on the berm? "Calculated guessing" will be our best approach to determine this amount.

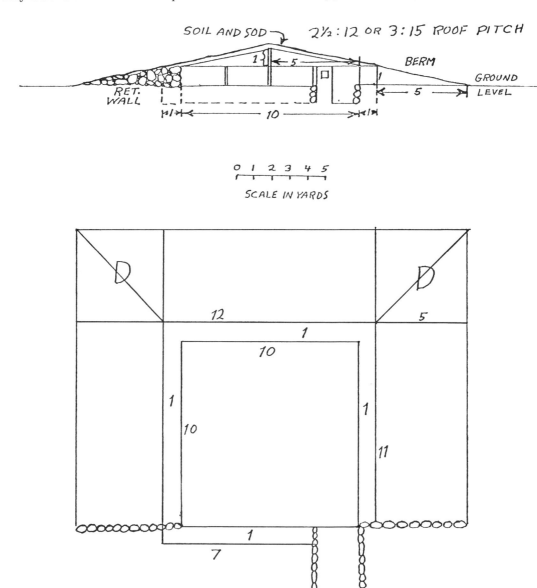

Illus. 3–6. Calculating the required depths of excavation for this hypothetical gabled, berm-style, 30' × 30' house should take into account poor percolation and the necessity for extra backfill.

The plan is for external wall dimensions of 30' square, or 10 yards by 10 yards. Let's say we were to excavate one yard (36") deep over the whole area within one yard of the walls. This square, 12 yards on a side and 1 yard deep, will yield 144 cubic yards of material ($12 \times 12 \times 1 = 144$). Because of poor percolation, the yard of space right next to the house won't be filled with the excavated earth. How much earth will the rest of the berm require? The three sides of the berm directly adjacent to the sand backfill have the cross-sectional shape of a right triangle: 1 yard high (h) and 5 yards wide (b); that is, a cross-sectional area of 2½ square yards ($A = \frac{1}{2}bh = \frac{1}{2} \times 5 \times 1 = 2\frac{1}{2}$). The total length of the berm is 34 yards ($11 + 12 + 11$) where it is directly adjacent to the sand backfill, so the volume is 2½ times 34, or 85 cubic yards. Add to this the volume of the two delta-wing shapes where the berms meet at the corners, marked "D" in Illus. 3–6. The volume formula for these corners is $\frac{1}{4}hb^2$, so, by substitution, $\frac{1}{4} \times 1 \times 5 \times 5 = 6\frac{1}{4}$ cubic yards. In all, it will require 97½ cubic yards of earth to build the berm ($85 + 6\frac{1}{4} + 6\frac{1}{4} = 97\frac{1}{2}$). But we've taken 144 cubic yards out of the hole! The 46½-cubic-yard difference is quite a bit to haul away or to spread around the site.

Before we make a second guess, let's consider the situation if the excavated material had been of good enough drainage to use for backfill. The backfilling can be considered as a rectangular volume 1 yard wide, 2 yards high, and 32 yards long ($10 + 1 + 10 + 1 + 10 = 32$). The formula for volume in a rectilinear solid ($V = lwh$) yields 64 cubic yards ($1 \times 2 \times 32 = 64$). In this case, the total volume of the berm right up to the walls is 161½ cubic yards ($97\frac{1}{2} + 64 = 161\frac{1}{2}$), a little more than the 144 cubic yards that came out of the hole. As loose earth occupies more space than it did originally, this isn't too bad. The berms could be made a little steeper, if necessary, or the excavation deepened very slightly. (As the hole gets deeper, the berms above grade get smaller.) Remember, too, that we'll need about 7 cubic yards of backfill along the front wall of this plan, which is similar to the original Log End Cave.

Let's assume that we have poor soil. Let's try excavating just 2½ feet instead of 3 feet. This time, the volume of the excavation will be 0.833 of what it had been before (because 2½ divided by 3 =

0.833), or 120 cubic yards ($0.833 \times 144 = 120$). The sand for the drainage remains the same 64 cubic yards. The berm is 6" higher now. The 3 in 15 pitch (3:15) established by the roof adds 30" to the width of the berm at original ground level.* Expressed in yards, then, the berm is 1.167 yards high and 5.833 yards wide. The volume of the berm, as we saw in the first calculation, is ½bhl (where l = length taken along the inside of the berm) plus $2 \times (\frac{1}{4}hb^2)$ for the two delta wing shapes marked "D." Substituting: $(.5)(5.833)(1.167)(34) + (2)(.25)(1.167)(5.833)(5.833) = 115.72 + 19.85 = 135.57$ cubic yards. We dug 120 compacted cubic yards out of the hole. Not bad. A bit deeper than 2½ feet (30") should be perfect.

The example is a realistic one, except for the obvious lack of a second entrance. Such a bermed house would be just slightly smaller than the Log End Cave we actually built. Perhaps a structure set only 30" deep shouldn't be called underground housing, but such a house is a closer relative to a subterranean house than it is to a conventional surface dwelling. The thick berm and earth-covered roof offer nearly the same advantages of heating and cooling as are enjoyed by an underground house. And the visual and environmental impact is about the same as that of the original Log End Cave. From a distance, the house would look like little more than a knoll on a flat landscape.

WASTEWATER SYSTEMS

At the earliest stages of planning and siting, consideration must be given to the disposal of wastewater. Wastewater disposal systems based on electric pumps are expensive beyond the budgets described in this book. Moreover, pumping systems are subject to ongoing maintenance, constant consumption of power over their lifetime,

---

*For this example, I've drawn the gable one yard (3') higher than the sidewalls, which are 5 yards (15') away from the midline of the house. The roof pitch, therefore, is 3' of rise for 15 lateral feet, expressed 3:15. Roof pitch is usually measured in terms of rise per 12 lateral feet. Our example is equivalent to a 2½:12 pitch. An earth roof should have a pitch of at least 1:12 to promote drainage, but not more than about 3:12, to avoid downward slumping of the earth. The Log End Cave pitch was 1¾:12.

reliability problems, and failure during power outages. Therefore, integrate the siting of a below-grade house very carefully with the location of the septic tank and drain field. On the side of a hill, waste drainage by gravity won't be a problem. On a flat site, it may be necessary to dig deeper tracks and drain fields than normal, keep the elevation of the house higher (which might mean more landscape sculpting to build the berms), or even raise the floor level of the bathroom by a step or two in order to establish a correct gradient for a gravity waste-disposal system.

If the water table at the building site is ever likely to be higher than the drains, rule out the site immediately. Rule out floodplains, too. Even if the walls are built as watertight as a swimming pool, the waste-disposal system would fail and probably back up into the home.

EXCAVATION

Together with backfilling and landscaping, excavation represents one of the biggest (and costliest) jobs connected with building underground. Estimates should be obtained from several heavy-equipment contractors. You can get an estimate for the whole job or you can pay by the hour for earth-moving equipment. If the contractor knows his business, the job estimate will be pretty close to the cost on a per-hour basis. Some contractors may tack on what appears to be a hefty profit for the job. You can't blame them. They also have to cover unforeseen circumstances when bidding by the job, such as overfastidious clients. I always get the per-hour charge, as well, and this is almost always the way I hire the equipment contractor.

Make sure that the contractors are pricing for the same thing; otherwise, no intelligent comparison can be made. For example, Contractor Smith gets $40 an hour for his backhoe, while Contractor Jones gets $45 an hour. But Smith uses an 18″ hoe, while Jones has a 24″ bucket. For excavation, Jones will come out cheaper, other things being equal. Ask the contractor if he charges for hauling the equipment to the site. Such changes can make quite a difference, especially when there's much follow-up work.

Bear in mind that you'll need equipment later for backfilling, landscaping, and perhaps a septic system. A contractor will be inclined to give a better price if justified by the volume of work. I stayed with the same contractor throughout the work at Log End Cave, and liked his work so much that I retained him again at Earthwood. I get excellent service as a regular customer.

Price isn't the only consideration; ability is another. If two contractors give similar estimates, but one has a better reputation, go with the good reputation, even if it costs a few dollars more. I also prefer to pay by the hour rather than by the job. There are so many imponderables in underground housing. You may change your plans during construction, about where to put a *soakaway* (dry well), for example. If you're paying by the job, the contractor will penalize you for changes in plans, and rightfully so.

WHAT EQUIPMENT IS NEEDED?

At the Cottage, our cellar hole was dug entirely with a backhoe. The backhoe worked well, so naturally I assumed that I needed one at the Cave, but the major excavation there was done much more efficiently with a front-end loader. The loader was $22 an hour (in 1977 dollars), the backhoe only $16, but the loader probably did the job in little more than half the time it would have taken the backhoe.

The excavation for the Cave was much bigger than the one for the Cottage and, because it was cut into a hillside, it was easy for the front-end loader, with its 6′ bucket, to maneuver. The entrance on the south side was the natural way for the loader to come in and out of the hole easily. It's true that the loader has to back away with each bucketful and dump it, while the backhoe can stand in one spot for a while and with its long boom place the earth outside the excavation. But the loader moves nearly one cubic yard with each scoop. Actually, to finish some of the corners, we did use a backhoe, instead; at that point, the loader had to travel too far with each load to dump it where it would be useful later for landscaping. It's a big plus to hire a contractor who has a variety of equipment, but make sure you're only being charged for one machine at a time.

The loader would also do well for the excavation of the bermed house used as an example in this chapter. In a relatively shallow excavation without a lot of big rocks, a good operator on a bulldozer can do a remarkable job quickly.

The backhoe is the only machine to use for digging the septic lines, soakaways, and drainage ditches, but wait until house construction is finished before embarking on these jobs. You want to be sure that grades are right (actual, not theoretical elevations are important here) and, too, you don't want a lot of dangerous ditches and holes all over the site.

4

The Footing

FOOTING DIMENSIONS

The footing, generally made of reinforced concrete, is the foundation base of the wall. The footing supports the entire structure and distributes the weight of the walls and roof over a base that's broader than the thickness of the walls. The footing is given tensile strength by the use of strong iron reinforcing bars, often called "rebars," so that the foundation takes on the characteristic of a monolithic ring beam. For relatively lightweight concrete-block construction, such as for a small block building, the dimensions of the footing follow a simple rule. The depth of the footing should be equal to the width of the wall. The width of the footing should be twice the width of the block wall it will support. Following this rule at Log End Cave, with its planned 12"-wide block walls, I decided on footings 12" thick and 24" wide.

A few years later, it was pointed out to me by concrete-foundation experts that using a 12"-thick footing is really "overkill." More than a 9" depth of footing is a waste of money on concrete. We're not building a skyscraper! Properly reinforced concrete is phenomenally strong on both compression and tension (the resistance against settling at a weak point in the subsoil). Earthwood is a much heavier home than Log End Cave, and is supported quite happily on a footing 9" deep and 24" wide. I still like the 24" width, because this decreases the load per square foot on the earth, resulting in a more stable building and less settling. Had I known in 1977 what I know now, I could have saved about a third of the money spent on concrete for the footings at the Cave. Let this be the first of several mistakes that we made which the reader can avoid.

SITE PREPARATION

The site should be excavated to a flat surface about two feet beyond the outside of the planned footing (Illustrations 4–1, 4–2). Check the level with a contractor's level and grade stick. Because of the slope that will be left if the work is done with a front-end loader, the 2' figure is consistent with the placement of the second set of white-birch pegs 4' outside those that mark the house corners.

On the flat excavation, mark the location of the four outside corners of the footing so that a backhoe can draw the tracks within which the footing will be poured. There are two ways to plot these corners: batter boards and educated guesswork. Going by the book, Jaki and I built four batter boards way up on the surface so that we could slide strings back and forth to make sure the sides were the right length and square. These boards can be seen in Illus. 4–2. A local contractor friend, Jonathan Cross, came over on a Saturday morning and found us struggling with these grotesque batter boards. We'd been at it for hours. We figured that once we had the batter boards established, we could use them for the footing's inside and outside dimensions as well as for the block walls. "Don't need 'em," said Jonathan. "They'll just get in the way of the backhoe." We moved to the educated-guesswork method then and there and it was much easier.

When you've got a levelled area with dimensions 4' greater than the dimensions of the footing, drive a 2 × 4 stake at the northwest corner (for example) 2' in from each side of the flat area. Put a nail in the top of the stake, leaving the nail head sticking out an inch for tying the mason's line. Buy a ball of good nylon mason's line; you'll be using it

Illustrations 4–1 and 4–2. Excavating a hillside with a front-end loader and its 6'-bucket. The site should be excavated to a flat surface about 2' beyond the outside of the planned footing.

frequently for this project. Measure the length of the footing along the north wall to a point (35'11" in our case), keeping about 2' in from the sloped edge of the excavation. Drive a stake into the ground and a nail into the stake. Now figure the hypotenuse (diagonal measurement) of your footing figures. We'll use our own figures as an example. Our footing dimensions are 30'8½" by 35'11" (30.71 × 35.92). Thanks to Pythagoras, we can calculate the hypotenuse (c):

$$c^2 = a^2 + b^2$$
$$c = \sqrt{a^2 + b^2}$$
$$c = \sqrt{(30.71)^2 + (35.92)^2}$$
$$c = \sqrt{943.10 + 1290.25}$$
$$c = \sqrt{2233.35}$$
$$c = 47.258' = 47'3''$$

These calculations can be done in a jiffy by using a calculator having the square-root function.

Now hook your tape to the nail on the northwest corner and, in the ground near the southwest cor-

ner, describe an arc with a radius equal to the shorter footing dimension, 30'8½" in our case. Next, hook the tape on the nail at the northeast corner and describe a second arc equal to the diagonal measurement, 47'3" in our example. The point where the two arcs intersect is the southwest corner. Drive in a stake and a nail there. Find the southeast corner by intersecting the east-side measurement with the south-side measurement. Check the work by measuring the other diagonal. The diagonals must be the same in order for the rectangle to have four square corners.

Perhaps the rectangle you've laid out doesn't use the cleared space to the best advantage. It might crowd one of the excavation slopes, but it might have plenty of room on the adjacent side. It doesn't take long to rotate the rectangle slightly to alleviate this problem. You might even have to do a little pick-and-shovel work if one of the sides doesn't have enough room.

"Calculated guessing" will take a few trials to get all four sides and the two diagonals to check, but it beats making batter boards that you're only going to use for one job. We actually had our four corners to within ½″ in twenty minutes, and that's accurate enough for the footing.

To get ready for the backhoe, place flags or white stakes on the various bankings for the operator to use as guides. You can set these guide stakes by eyeballing. Sight from one stake to another and instruct a helper to plant a third marker on the banking in line with the two stakes. In all, you'll place eight guide markers.

TRENCH DEPTH

How deep should you dig the footing trench? Figure this depth carefully. The important relationship is the one between the level of the top of the footing and the top of the floor. The cross-hatching in Illus. 4–3 represents undisturbed earth. For underfloor drainage, bring in 4″ of sand to lay the floor on.

I like a 4″-thick concrete floor, but if you've got a good honest 3″ at the thinnest portions, you've got a good strong floor. The advantage of a full 4″ is simply more thermal mass for heat storage. The exception to the 4″ floor would be if you intend to incorporate an in-slab heating system, such as a forced-hot-water system in rubber tubes laid in the slab itself. In this case, a 5½″-thick floor is recommended by manufacturers of such systems. People who have in-slab heating speak very highly of it. Get information on products from your local plumbing-and-heating supply store. Be sure to insulate with extruded polystyrene below the slab. How much? At least an inch, but check with the supplier for the requirements of your locality.

Illus. 4–3 shows that the concrete floor is designed to resist lateral pressures on the base of the wall and the possibility of the first course of blocks breaking loose from the footing. Although the actual Cave footings were 12″ × 24″, the drawing shows a footing with a sufficient 9″ depth. Two inches of the concrete floor and 4″ of compacted sand will be below the top of the footing. This leaves just 3″ of additional material to be excavated to accommodate the footing track. This footing track should be square, level, and about 30″ wide to easily accommodate the 24″ footing, two forming boards, and 2″ of extruded polystyrene insulation. Digging the footing track is probably best done by hand, unless digging conditions are poor because of hard soils or large boulders. In such cases, a backhoe can save a backache.

Illus. 4–3. Footing details

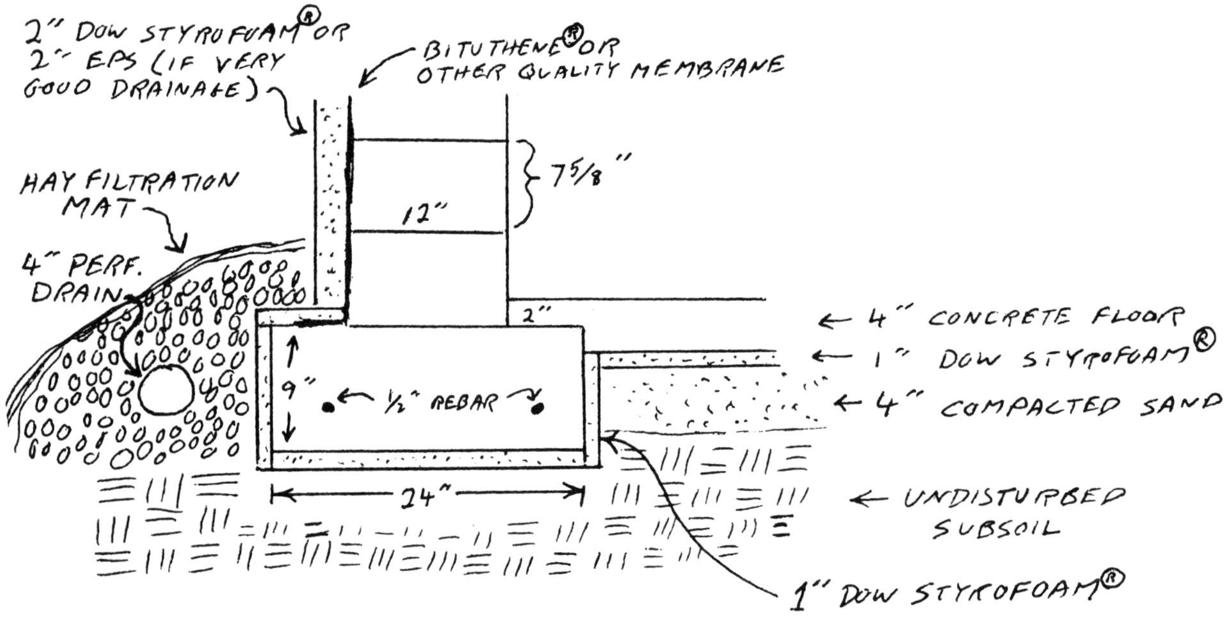

KEYED JOINT

There's another way of "keying" the first course of blocks to the foundation. A keyed joint can be created by setting a piece of wood flush with the surface of the wet concrete, and halfway between the inner and outer forming boards. A good keyway can be made by rip-sawing a 2 × 4 down the center, using a circular saw. If the blade angle of the saw is set at about 75° instead of 90°, an excellent *draft angle* will be created so that the board can be easily removed from the partially set concrete. Oiling the keyway board is also strongly recommended. The resulting keyed joint will look like the one shown in Illus. 4–4. Note that the draft

below the floor instead of 4". Note that there's also 1" of extruded polystyrene (such as Dow Chemical's Styrofoam® Blueboard®) under both the footing and under the floor.

INSULATING THE FOOTING

Wrapping the footings with extruded polystyrene is very important. At Log End Cave, we didn't include this detail and suffered condensation at the base of the wall and about 6" in from the floor during warm moist conditions in May, June, and early July, when the footing was still conducting

Illus. 4–4. A keyed joint is created when an oiled keyway board is removed from the concrete after it's set. The draft angle is kept to the outside of the footing.

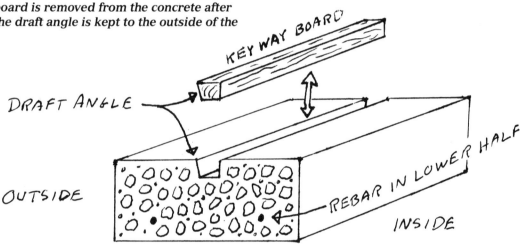

angle is kept to the outside of the footing. Later, the first course of blocks can be firmly tied to the footing by filling the block cores halfway with concrete. If I were doing a rectilinear earth-sheltered house again, this is the method I'd choose, because the blocks wouldn't be in the way of *screeding* (flattening the top of the concrete so that it's even with the top of the forming boards) and finishing the floor, as they were at Log End Cave. Also, using this method, it's really pointless to excavate footing tracks at all. The forms can be built right on the floor of the excavation, providing the floor is flat and level.

Another method of eliminating footing-track excavation is to set up footing forms right on the flat excavated site and use 7" of compacted sand

"coolth" from the still-cold soils at 7' of depth. Not until the footings warmed up near the end of July did the condensation disappear. Wrapping the footings (and floor) with extruded polystyrene stops this condensation, as we've proven at Earthwood, where there's no condensation at any time of the year. The Blueboard® used at Earthwood keeps the footing temperature above the dew point. Illus. 4–5 shows the difference.

Under the footings, it's important to use genuine Styrofoam® Blueboard®, which has a compression strength of 5600 pounds per square foot with only 10% deflection. At most, the load per square foot on a Log End Cave type house will be less than half that, so deflection (compression) of the Blueboard® will be considerably less than 10%.

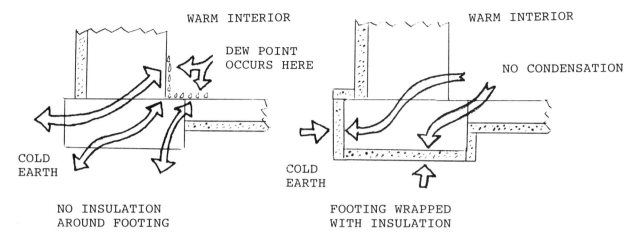

Illus. 4–5. At Log End Cave (left), the footing conducts heat to the earth. Low footing temperatures in early summer can cause condensation. At Earthwood (right), insulation keeps footing temperatures above dew point, and the concrete stays dry.

I don't know of any other extruded polystyrene foam with sufficient compression strength. Do *not* use *expanded* polystyrene (also known as *bead-board*) around the footings or under the floor.

A final word about the footing tracks: If a large boulder has to be removed, be sure to fill the hole with well-compacted earth or sand. Be sure to wet such material for maximum compaction. Concrete should always be poured over undisturbed earth. If the subsoil is disturbed, mechanical compaction is imperative.

FROST WALL

In northern areas, it's necessary to prevent footings close to the surface from *frost heaving*. Heaving can occur when wet ground beneath the footing freezes and expands, pushing upwards on the footing. The massive weight of the wall is of little help. The expansion forces of freezing don't care how much weight they're called upon to lift. The only solution is to make sure that the ground below the footing won't freeze. This can be done by going deeper with the footing, protecting it with extruded-polystyrene insulation, or both.

The frost wall at the original Log End Cave is that portion of the south wall which is entirely above final grade, 8' of wall near to and including

the door. The footing below this portion of wall doesn't have the protection of 3' of earth, as does the rest of the south-facing wall. So we decided to increase the footing depth to 24" over a 12' section in the southwest corner. This extra concrete worked fine; the foundation has never heaved in 17 years. In theory, the frost wall should have descended to 48", the depth of frost penetration in our area. But this protected south-facing wall doesn't experience that extent of frost penetration. Extra care would be imperative for unprotected footings on any other wall.

There's another method of protecting against frost heaving which is being used more and more frequently. The method uses insulation to take advantage of frost's tendency to permeate into the soil from above at a 45° angle. Illus. 4–6 shows how this works. The advantages of this system are ease of construction and savings on materials.

Extend a minimum of 2" of extruded polystyrene (R-10) away from the frost footing to a distance equal to the local frost depth. This protects the cross-hatched area shown in Illus. 4–6. The rigid foam is protected from ultraviolet radiation by covering it with 6-mil black polyethylene and 3" of #2 crushed stone. Pitch the foam and poly away from the home for drainage purposes. Alternative methods of frost protection, shown in the insets, are very good, but they involve more excavation and replacement of soils.

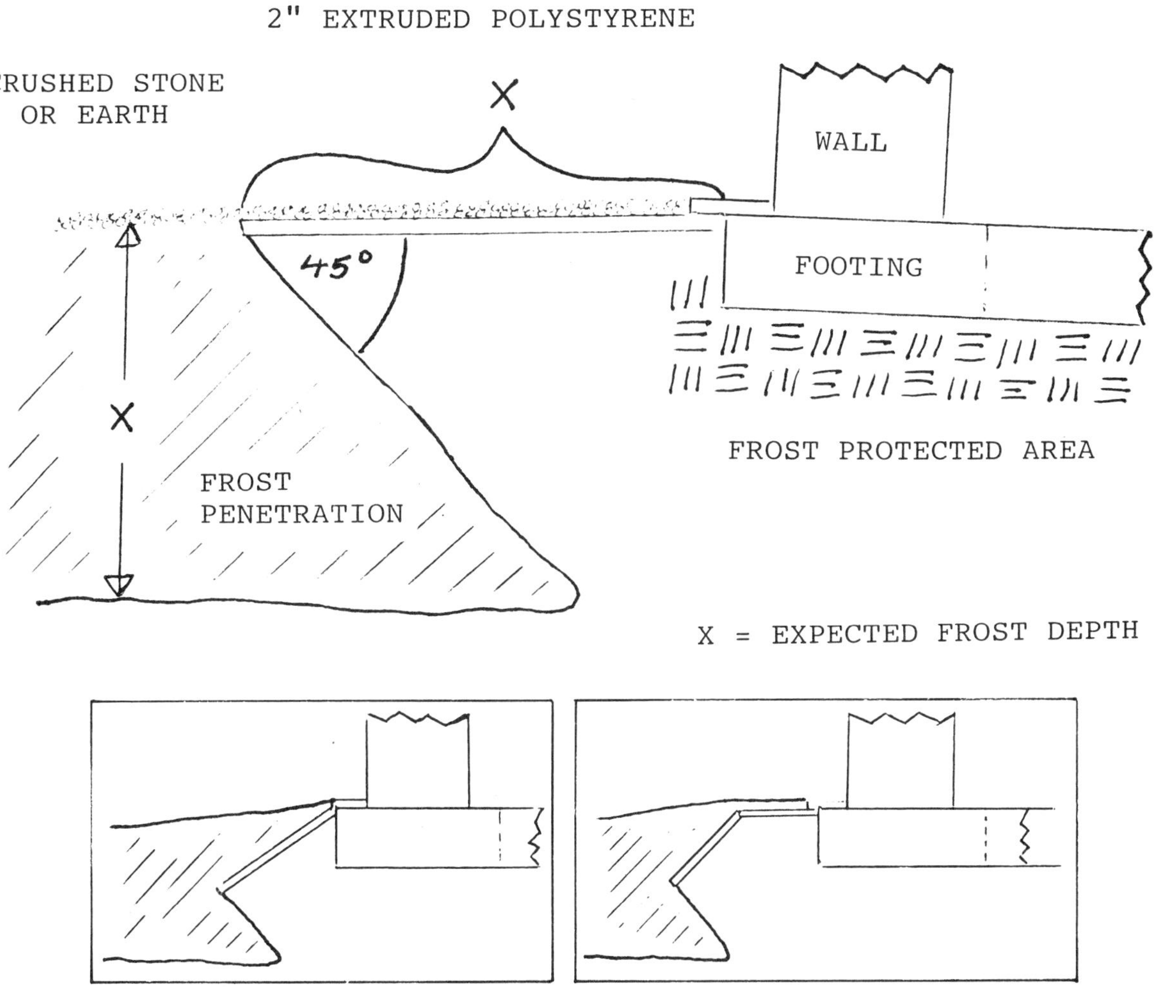

2" EXTRUDED POLYSTYRENE

CRUSHED STONE
OR EARTH

WALL

FOOTING

45°

X

FROST
PENETRATION

FROST PROTECTED AREA

X = EXPECTED FROST DEPTH

Illus. 4–6. Because frost penetrates on a 45° angle from the surface, footings can be protected by the placement of rigid-foam insulation. The insets show alternate methods of placement.

FORMING

We used 2″ × 10″ forming boards kindly lent to us by Jonathan Cross. These worked out in forming for our 12″-deep footing (we just let the extra 2″ concrete at the bottom take the shape of the track itself), but they would have been perfect for making a 9″-deep footing with an inch of Styrofoam® at the bottom of the form.

Preparing the forms takes time. First, figure out exactly what lengths your forms need to be. No room for error here; you don't want to cut a good plank wrong. I drew a diagram of the whole forming system, showing clearly how the corners were to be constructed (Illus. 4–7). The long forms on

the north and south sides are 3″ longer than the footing, to allow the 1½″-thick east- and west-side forms to butt against them. Similarly, the east and west forms on the inner ring are 3″ shorter than the inner measurement of the footing because the thickness of the planks of the other two forms will make up the difference. All of the other forming boards are consistent with the actual footing measurements. (The footing measurements given are based on the location of the 12″ blocks laid up by the surface-bonding method.)

It will be necessary to cleat planks together (Illus. 4–8) to make each full length of the form. Cut cleats at least 3′ long and use 16-penny scaffolding nails to fasten the cleats to the planks. Scaffolding

OUTSIDE FOOTING DIMENSIONS: 35'11" × 30'8½"

INSIDE FOOTING DIMENSIONS: 31'11" × 26'8½"

DRAWING NOT TO SCALE

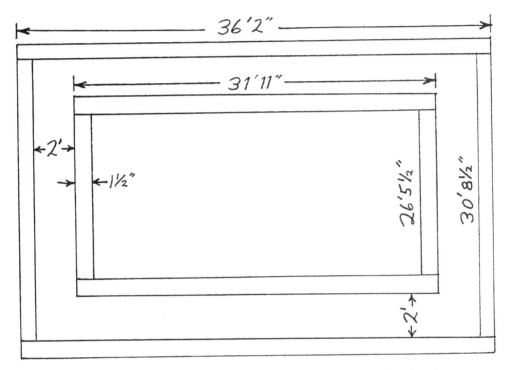

Illus. 4–7. Forming diagram showing the lengths of all footing forms

4-8

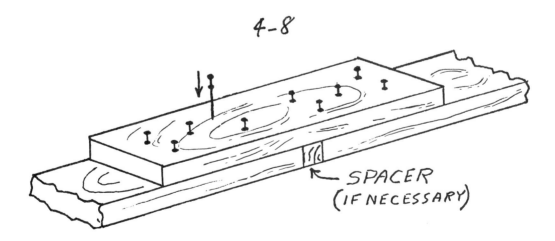

Illus. 4–8. Cleating together two planks

(or *duplex*) nails have two heads so that the nails can be easily removed when the forms are dismantled. I use ten nails on each cleat, five for each plank. Butt the planks tightly together and eyeball them dead straight with one another before nailing the cleat. Note that the required measurement of our north and south exterior forms is 36'2". To make good use of Cross' 18' planks, we cheated a little and left them 2" short of butting against each other. This is okay as long as you install a little spacer (Illus. 4–8) into the gap. The spacer will give almost the same stability as two planks butting directly against each other.

You'll need a lot of strong 2 × 4 stakes, about 24" long. Allow five or six for each length of forming and a few extra for the ones you'll smash to splinters with the sledgehammer. I was fortunate in borrowing stakes from Jonathan, but you can make stakes from economy-grade studs. Fifteen 8-foot 2 × 4s will make sixty 24" stakes. Put well-tapered points on each stake.

PLACING THE FORMS

This job would be difficult without a contractor's level. Beg, borrow, or rent one. The most important consideration in setting up the forms is that they be level with each other. Set up the level at some point outside the foundation where you'll have a clear view of a grade stick held at each of the four corners (Illus. 4–9). Using the existing corner stakes as guides, bring in the longer of the north-side forms and put it roughly in place. Drive new stakes into the ground at the northwest and northeast corners, positioning them so that they'll be on the outside of the forms. Don't drive stakes on the side of the forming boards where the concrete is poured. Drive all the new corner stakes so that they're at the same elevation, as judged by the contractor's level. This will be about 10" above the average grade of the excavation or of the footing tracks, depending on the relationship of the floor

Illus. 4–9. The most important consideration when placing the footing forms is that they be level with one another.

level to the footing. This relationship, in turn, depends on which method you've chosen for keying the first course of blocks. Make a cross-sectional sketch (similar to Illus. 4–3) of this relationship, but including all of the details of your floor: footing dimensions, insulation, keying method, etc.

To get the average grade of either the foundation area in general or of the footing tracks, take 12 readings at equally spaced locations, and average the results.

Nail the long form to the new stakes so that the top of the form is level with the top of the stake. Eyeball the form straight and drive a third stake into the ground about halfway along. Only the corner stakes need to be at the same grade as the top of the form, so that they can serve as benchmarks to work from. Other stakes can be driven slightly lower than the level of the forms, so that they will be out of the way of screeding when the pour is made.

Now level the form. This chore is easiest with three people: one to hold the grade stick, one to read the contractor's level, and one to pound the nails. Again, use scaffolding nails, coming in from the outside of the stake and into the forms. Drive the nails into the stake before you pound the stake

into the ground, so that the point of the nail is just barely showing through the stake. Use a rock or sledgehammer to help resist the pressure as you drive the nails all the way into the forms. Put at least one stake between the corners and the midpoint, maybe two over an exceptionally long span. Check for level along the whole form.

One down, seven to go. The rest are done in the same way as the first. Complete the outer form before proceeding to the inner ring. Make sure the diagonals check! The inner ring is placed to leave a space equal to the width of the footing, 2' at Log End Cave. Make 26"-wide tracks to allow for the inch of extruded polystyrene on each side of the concrete.

Make the inner ring level with the outer ring. Check the whole job with the contractor's level, moving every 8' or so around the forms. Make slight adjustments by moving the stakes up or down. Use a lever to move stakes upwards, a sledge to move them downwards. Hit the stakes, not the forms. It may be necessary to clear some earth from beneath the forming boards to allow them to settle into the proper grade. Finally, nail 2 × 4 buttresses every 10' or so to resist the tremendous outward pressure exerted on the forms

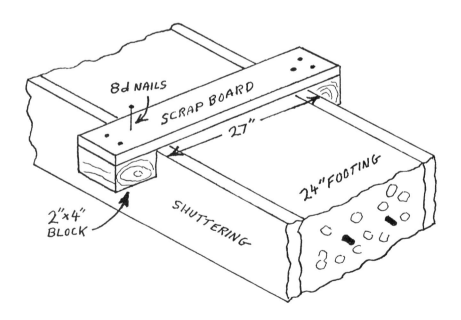

Illus. 4–10. Movable cleats will keep the forms (shuttering) intact during the pour.

during the pour. These buttresses can be footed back to the walls of the excavation. Another means of resisting this pressure is to construct about a dozen movable cleats (Illus. 4–10).

FINAL PREPARATIONS BEFORE THE POUR

With the forms firmly in place, it's time to install the very important rigid-foam insulation. As previously stated, the rigid foam at the bottom of the footing should be Blueboard® for its compression strength. Other *extruded* polystyrenes (not *expanded*) can be used along the sides, such as Dow's Grayboard®, but you might as well use the Blueboard® for all purposes, as you need it under the footing, unless there's a significant cost savings with one of the other extruded polystyrenes. An added advantage of using the rigid foam is that you won't have to oil the forms for removal. On buildings with no floor dimension greater than 40′, expansion joints aren't needed, and you could just put the insulation 4″ up from the bottom on the inside of the forms. Old engine oil painted on to the top 4″ of the inner forms will facilitate form removal. If floor dimensions exceed 40′, just leave the rigid foam right to the top of the inner forms and the foam becomes your expansion joint.

Reinforce the footings with ½″ rebar, two bars throughout the footing. (Some building inspectors may require three pieces, but two is really sufficient for the tensile strength required.) The rule of thumb for overlapping reinforcing rod is that joining pieces should be overlapped by 40 times the diameter of the rebar. So the overlap with ½″ rebar should be about 20″. Tie the overlapping pieces together with forming wire. Bend right angles in some of the pieces for use at the corners.

The rebar should be placed in the bottom half of the footing pour, in order for it to lend its maximum tensile strength against settling. This positioning is shown in Illus. 4–3. The rebar can be supported during the pour by pieces of broken bricks or small 3″-thick flat stones. Special wire supports made for the purpose, called *frogs* or *chairs*, are commercially available. Whichever

supports you use, they'll become a part of the concrete, so choose something clean, strong, and of the right height, about 3″. Keep the rebar at least 3″ in from the edges of the forming track.

We used ⅝″-thick iron silo hoops for rebar at both Log End Cave and Earthwood. These hoops were left over when we recycled fallen or leaning silos for their spruce planking. If someone has torn down a silo in your area, try to work a deal for the hoops; they make strong, cheap rebar. Old hoops usually have to be cut apart with a hacksaw, and straightening them can be a job, but the savings can be worth the effort. Yet, construction rebar isn't very expensive, so weigh time and effort against money.

As accurately as possible, calculate the amount of concrete you need. It sells by the cubic yard and the going rate in 1994 in northern New York was $60/cubic yard for 3000-pound-test mix, which is what you should be ordering. Here are the calculations for the Log End Cave footings. If you follow this example, you'll be able to do your own calculations for footings of different dimensions.

The volume of concrete equals the cross-sectional area of the footing times the perimeter. $V = 1′ \times 2′ \times 126′ = 252$ cubic feet. Dividing by 27 gives cubic yards: $^{252}\!/_{27} = 9.33$ cubic yards. We added an extra cubic yard for our deep frost-wall footings at the southeast corner, which gave us 10.33 cubic yards. Actually, our computation was a little more involved than this, since the footings actually measured about 11″ × 23″ in cross-section. This adjustment reduced the figure to 9.3 cubic yards. The cement truck had a maximum capacity of 9.5 yards, very close to our estimate, so I called for a full load. We ended up with about a wheelbarrow of concrete left over.

It's better not to cut your estimate quite that close. Have a place to put any leftover concrete, some small project. Form a sidewalk, a trashcan pad, a playhouse foundation, anything to make use of any excess. You can even make concrete paving slabs, always useful for a variety of purposes.

On the morning of the pour, apply a coat of used engine oil to any part of the forms in contact with the concrete. This coating will make removing and cleaning the forms much easier. If you borrowed the forms, you'll want to return them in as good a condition as you found them.

POURING THE FOOTING

Round up plenty of help on the day of the pour. If they're fairly sturdy individuals, a crew of four or five is sufficient to draw the concrete around the ring (Illus. 4–11). Have a few strong garden rakes or hoes on hand. A metal rake is the best tool for pulling concrete along between the forms.

You'll be charged for overtime if the cement-truck driver has to stand around and wait for last-minute preparations. Have all the forms set and supported, and the insulation and rebar in place. Make sure access to the site is good. Have a wheelbarrow and planks ready to wheel concrete to awkward spots. Ideally, the driver should be able to chute the concrete to any place in the forming.

Ask the driver for a stiff mix; a 3" *slump* is good, should he ask you for such a figure. Slump refers to how much the concrete sags when a test cone is removed from a fresh sample. The more water the easier the work, but the weaker the concrete. Soupy concrete can't be the 3000-pound-test you ordered and might be only half as strong.

There's not much to say about drawing the concrete around the forms except that it's darned hard work, especially stiff concrete. In order to prevent voids, try to vibrate the concrete down into the forms with your tools. After one side has been poured, one or two folks can start to screed the concrete before it sets too much. Screed with a 2 × 4 (Illus. 4–12), drawing the concrete along with a constant backward and forward sliding motion. You can use the screed board on edge for the initial flattening, and then for a relatively smooth surface give it another screeding with the side of the 2 × 4. If you intend to use the keyway method shown in Illus. 4–4, now's the time to insert the oiled keyway pieces. Push them right into the concrete in the center of the track and screed over the pieces, leaving the top wooden edge showing on the surface.

Can you home-mix concrete for the footings or the floor? I won't mix my own concrete. The savings aren't really very great, especially if any value at all is placed on your time. And a footing poured over several days won't be as strong, because of all the "cold joints" between sections.

Clean all your tools at the conclusion of the pour. The cement-truck driver should have a spray

Illus. 4–11. A crew of four or five, each with a good strong rake, should be sufficient to draw the concrete around the footing ring.

hose right on board which he uses to clean his chutes, and he'll let you spray off your rakes, shovels, hoes, and wheelbarrow.

REMOVING THE FORMS

Concrete dries at varying rates depending on its strength, stiffness, the air temperature, and the humidity. You can usually remove the forms the next day, and they should be easy to remove if you've included the 1" rigid-foam buffer up against the forming boards. Clean the forms and stakes for reuse, particularly if they belong to someone else.

It took six of us two hours to draw and level the concrete. This is as it should be. You don't want to pay overtime for the cement truck. Jaki and I removed the forms the next day, breaking one which was held fast at a place where the concrete had leaked.

Illus. 4–12. The concrete can be screeded (made level with the top of the forming) with 2 × 4s.

5
The Floor

PREPARATION

The preparation for each job is usually more work than the job itself. We poured the footing on July 11th. Although we'd hoped to be ready for the floor pour by the 16th, we weren't fully prepared until July 23rd. It's incredible how many things there are to do before the floor can be poured.

Contractors will often pour the footing and the floor together on the same day. This concrete mass is known as a *monolithic slab*. The footing portion, sometimes referred to as a *thickened edge,* is merely a thicker part of the floor pour. A 4″ floor with a thickened edge of 8″ or 9″ deep and 16″ to 24″ wide is common. In fact, we poured both our summer-cottage slab and my workshop slab at Earthwood monolithically. But I don't recommend this method to first-time builders on a full-size house foundation. The task should be broken down into two more manageable bites, footing one day, floor a different day. The kinds of preparation are different, too, and there's a great advantage to using the completed footing as a sturdy and uniform edge for screeding the floor, an advantage you won't have if you pour a monolithic slab.

Because of our plan of footing the base of the block wall with the top 2″ of the floor pour, it was necessary to lay the first course of blocks all the way around the building. If I were to build a similar building today, I'd choose the keyway system shown in Illus. 4–4, but in this narrative, I'll follow what we actually did.

For good under-floor drainage, and before doing the block work, we spread 4″ of sand over the floor. By July 14th, the footing was aged enough to build a ramp of sand over the frost-wall portion. We brought in two loads of sand and dumped them just outside the "door." The bulldozer was then able to push and backblade the sand onto the floor. Using the footing as a guide for levelling, the operator did a good job of spreading the sand evenly over the area. He'd done all he could in an hour, and we took our time finishing the job by hand, checking with the contractor's level now and then. The next day I hired a power compactor and tamped the sand, with Jaki watering the area between tampings, using a watering can, since we had no running water on-site. This power compacting is very important to give the floor a solid base. Tamping dry sand is an exercise in futility, so be sure to soak the sand between tampings. I like to hit each square foot at least three times with the compactor.

SURFACE BONDING

Our 12″ concrete block walls are laid up by the *surface bonding* method. In brief, the blocks are dry-stacked without mortar, and a strong stress-skin coating of *surface-bonding cement* is applied to each side of the wall, bonding block to block. This cement coating is saturated with millions of tiny glass fibres, which impart a tremendous tensile strength to the wall. The tensile strength of this monolithic membrane applied to the wall has been found to be six times stronger than the tensile strength of a conventionally mortared wall. Details of construction are given in the next chapter, but laying up the first course will be discussed here, because a completed first course was necessary to pouring the floor.

With surface-bonded blocks, the first course is laid on the footing in an almost conventional manner: on a ⅜″ to ½″ mortar joint. This mortar's

Table 2
Dimensions of Walls and Wall Openings Constructed with Surface-Bonded Concrete Blocks*

Number of blocks	Length of wall or width of door and window openings**	Number of courses	Height of wall or height of door and window openings***
1	1' 3⅝"	1	7⅝"
2	2' 7¼"	2	1' 3¼"
3	3' 10⅞"	3	1' 10⅞"
4	5' 2½"	4	2' 6½"
5	6' 6⅛"	5	3' 2⅛"
6	7' 9¾"	6	3' 9¾"
7	9' 1⅜"	7	4' 5⅜"
8	10' 5"	8	5' 1"
9	11' 8⅝"	9	5' 8⅝"
10	13' 0¼"	10	6' 4¼"
11	14' 3⅞"	11	6' 11⅞"
12	15' 7½"	12	7' 7½"
13	16' 11⅛"	13	8' 3⅛"
14	18' 2¾"	14	8' 10¾"
15	19' 6⅜"	15	9' 6⅜"

*Standard 16-inch blocks, 15⅝ inches long by 7⅝ inches high.

**Add one-fourth inch for each approximate 10 feet of wall to allow for nonuniformity in size of blocks.

***Make a trial stacking of blocks to determine the actual height of wall or opening before beginning construction.

purpose isn't to "glue" the wall to the foundation, but rather to accurately level the first course for the later placement of dry-stacked blocks. The only difference between the treatment of the first course of surface-bonded blocks and the first course of blocks in a "normal" wall is that in a surface-bonded wall no mortar is placed between adjacent blocks; the blocks are butted tight against one another.

The footing dimensions of Log End Cave accommodate full courses of blocks without the need for cutting any blocks. When using surface bonding, the true size of the blocks must be used in calculations. Standard 16" blocks, for example, area actually 15⅝" long by 7⅝" high. Table 2 gives true dry-stacked wall dimensions.

Note that ¼" is added for each 10' of wall to allow for nonuniformity of blocks. Our block plans were based on size and space requirements of our general floor plan, as well as the availability and dimensions of certain materials, especially the roof-framing timbers. For greater lateral stability, we chose 12″-wide blocks, instead of 8″ or 10″ blocks. The east-west length of the Cave, then, is 26 whole blocks and one block laid widthwise, shown in Illus. 2–2. We can break 26 down to 13 blocks twice and use the chart: 16′11⅛″ × 2 = 33′10¼″. Add 12″ for the block laid widthwise (this is the block that "turns the corner") and add ¾″ to allow for the nonuniformity of the blocks, and a total length of 34′11″ is obtained. Similarly, the east and west walls can be computed: 14′3⅞″ (11 blocks) + 14′3⅞″ (11 blocks) + 12″ (the turned block) + ¾″ (nonuniformity margin) = 29′8½″. The south wall has the same number of blocks as the north wall, except that five blocks are left out where the door panel is located.

BLOCKS

The first task in laying the first course of blocks is to mark the corners. Use methods similar to those already described for establishing the corners of the footing. In theory, the outside edge of the block wall will be 6″ in from each edge of the footing, but it's advisable to obtain ¼″ accuracy, so fiddle with the tape and your marks until all the walls are the right length and the diagonals check. Now snap a chalkline on the footing between corners marking the outside of the wall. Stick to your block plan to get the first course of blocks in the right place. It's necessary to cut one 4″ × 8″ × 16″ solid block down to a 12″ length to complete this and every course. By using a hammer and cold chisel, it's

easy to cut 4″ off a 4 × 8 × 16 solid. Mark the block with a pencil and score the pencil line (both sides of the block) with hammer and chisel. Go around the block again, hitting the chisel a little harder. Before long, the block will break right along the scored mark.

Let's discuss blocks: the various sizes and materials, how they fit together, cost, and so on. We chose 12″-wide concrete blocks: 12″ for a stable wall, concrete for strength. Cinder blocks are lighter, cheaper and easier to use, but they don't compare with concrete for strength. The blocks we used have three full cores and two half-cores in them, except for the corner blocks, which have the space filled where the half-cores would normally be (Illus. 5–1). The regular block weighs 60 to 65 pounds, corner blocks about 5 pounds more. You could use 8″ blocks, but with these it's necessary to build a pilaster every 10′ around the perimeter of the wall (Illus. 5–2) to resist lateral pressure.

We used 8″ blocks and pilasters in the basement of our first home, Log End Cottage, and we grew to dislike the pilasters, which got in the way of the floor plan. They're also a pain to apply surface bonding to. They slow production enormously, both when laying them and when surface-bonding. We decided to use stable 12″ blocks and we filled the cores with vertical rebar and concrete at every location that would normally take a pilaster.

Ten-inch blocks are offered by most block companies, but I can't recommend them. Their awk-

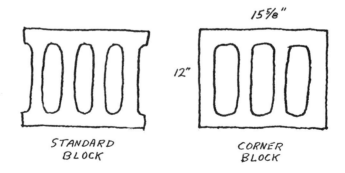

Illus. 5–1. Standard 12″ concrete blocks, weighing about 65 lbs. each, have half-cores at the ends; corner blocks are filled at the ends and weigh about 70 lbs.

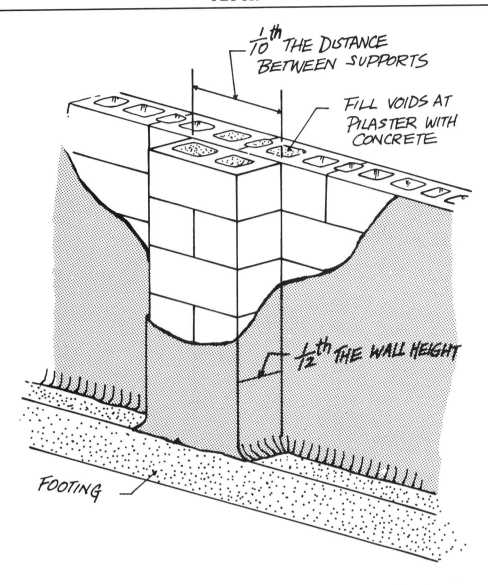

*Illus. 5–2. A typical pilaster for strengthening a wall or supporting a beam.
Use this pilaster if you'll be using 8″ concrete blocks.*

ward dimensions mean constant cutting of blocks, and you still need those pesky pilasters. All blocks, by the way, are 15⅝″ long by 7⅝″ high.

Uniform blocks made especially for surface bonding are available in some areas. The blocks are passed through surface grinders at the block plant, assuring uniformity. They are, of course, more expensive than "normal" blocks.

To avoid cutting blocks, 12″ blocks should be used with a block plan similar to our own. Cutting blocks, especially cored blocks, is difficult. A little adjustment to the dimensions of your house can save time. We end each course with a block turned widthwise (Illus. 5–3), instead of having a whole number of blocks on a side.

Ask for a volume discount. You could set yourself up as a contractor and open a charge account with the block plant; this approach could save you

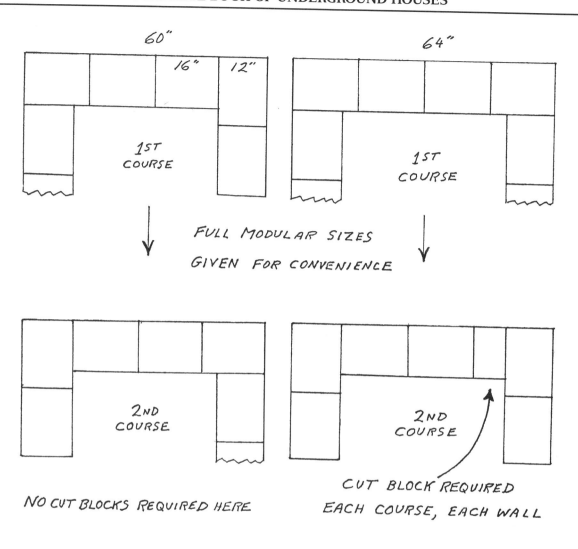

Illus. 5–3. Adjust your plans a bit so that corners can be made without having to cut blocks.

money. You should get at least 10% off for 500 or more blocks. Ask about haulage charges, too. These charges usually vary with distance, but they might also be a function of volume. If you're fortunate enough to have two or more suppliers, get comparative bids. Compare the blocks. With surface bonding, it's worth a few cents extra for uniform blocks. Poorly made blocks at $0.10 off are no bargain. In 1977, we paid $0.51 per block; in 1994 the same block cost $1.10.

LAYING THE FIRST COURSE

Lay the first course of blocks in a rich mortar of 1 part masonry cement, 1 part Portland cement, 5 parts sand, equal parts by volume. Make a stiff mix, as for stone masonry, particularly if the footing is uneven and the mortar has to be thick in places.

Start with the highest corner, but know how much higher it is than other parts of the wall, so you can estimate a practical thickness for the mortar bed. Remember that the purpose of the first course is to establish a level base for the dry-stacked courses, so the thickness of the mortar bed may vary around the footing. Set up the four corner blocks so that they're all precisely the same level. Then tightly stretch a nylon mason's line

allowed quite a lean to develop in the first course, and it took considerable shimming later on to correct it.

Keep the blocks tight to each other (Illus. 5–4). Make sure no mortar has crept in from below between the blocks. A light tap with a hammer will usually assure that two adjacent blocks are well butted against each other. When you lay the last block, you may find that it doesn't fill the last gap exactly. You may have to make small adjustments, even opening tiny spaces (less than 1/8″) between blocks. I found it useful to check the length of the

Illus. 5–4. Keep the blocks tight against each other.

from the top of one corner to the next. Stretch the line as tight as you can and tie it to something that won't move, like another 12″ block. Eyeball the line. There will probably be a 1/16″ dip. Keep that in mind as you proceed along the course, so that halfway along, the block is, say, a string's width above the line. Check each block for level as you lay it. The string will tell you the level along the length of the wall, but use a two-foot level to check along the wall's width. This checking is important. Laziness here will cost you time later. On our north wall, I

first six blocks I laid, to find out how my work compared to the values given in Table 2. Any adjustments to the quality of the work could then be made. Remember that we allowed an extra 1/4″ for each 1′ of wall. I found this to be a generous allowance, but another company's blocks might be slightly different. We set a 4″ × 12″ block (cut from a 4″ solid) inside the last full-size block on the south wall to establish the base for a short pillar, which strengthens the end of that wall. This can be seen in the block plans in Illustrations 2-2 or 2-5.

I'd never laid blocks before, though I'd worked for masons, but it only took about ten hours to get a decent job, decent except for the aforementioned north wall, where I didn't use the level often enough.

This first course of blocks established the edge of the concrete floor at Log End Cave, and provided a flat level surface for screeding the floor. No further blocks were laid until the floor was finished. In fact, the main part of the block order wasn't even delivered until the floor was hard, so that the pallets of blocks could be set down on the floor for convenient access during wall building. We hauled the first 90 blocks for the first course with our little pickup truck, three heavy loads. Again, consider strongly the keyway detail already described. It's easier to pour the floor to the same level as the footings, and all the blocks can be delivered to the site in one load.

PILLAR FOOTINGS

The three main girders at the Cave are well supported by posts cut from 8"-square and 9"-square barn timbers. Each of these posts must have a pillar footing beneath it to support the heavy roof load without cracking the floor. We decided to make the footing under each post 24" square by 12" deep, although 9" deep would have been sufficient. We measured for the location of each footing, scaling off the rafter-framing plan, and with a stick marked the squares in the sand. Because the level of sand was already designed to carry a 4" floor, we only had to excavate an additional 8" to give us the 12" depth. Pick and shovel work. We hauled the material away with a wheelbarrow, although we used some of the sand to build up one or two low spots on the sand pad. It was stiflingly hot the week we prepared for the floor pour, and the dogs found these pillar-footing holes (Illus. 5–5) to be ideal spots for lapping up some of the cooler earth temperatures below the sand.

Although at the Cave we poured these pillar footings at the same time as we poured the floor, we varied a few details three years later when we built Earthwood. There, we poured the lower half of the 9"-deep pillar footings at the same time that we poured the perimeter footing, and we included an inch of Styrofoam® Blueboard® under the pour

Illus. 5–5. The bottom half of the pillar footings can be poured at the same time as you pour the floor.

Illus. 5–6. The bottom half of the pillar footings are poured and the concrete is left rough to make a good "cold joint."

Illus. 5–7. Styrofoam® and wire mesh is in place. There's already Styrofoam® under the pillar footing.

to stop much of the energy "nosebleed" to the earth. This meant that when we poured the floor a few days later, we didn't have these awkward holes underfoot to trip in (Illustrations 5–6 and 5–7).

UNDER-FLOOR DRAINS

The sand under the floor provides drainage for water which might find its way under the floor for any reason. To carry the water away, install 4" perforated drain tubing (also called *drain tile*) in this drainage layer, and slope the tubing away to some point downgrade from the site. My method, which we used again at Earthwood, is to use a hoe to draw a track in the sand for the drain tubing, spiralling around the sand pad in such a way that no point under the floor is more than a few feet from the drain tile. T-connectors are made for joining branch lines to a main line. I try to slope the drain ever so slightly towards the point where it exits the floor under the footing. Outside the foundation, use nonperforated tubing, and continue it

away from the site, either to a large soakaway hole filled with stones or, better, to a point out above grade. The exposed end of the drain tubing should be covered with a vermin-proof screen, such as ½" mesh hardware cloth. The drain tubing I use is a black, ribbed, flexible pipe with a white nylon or fiberglass filtration sock to keep silts out of the pipe. This tubing is cheap insurance against water problems from below. See Illus. 5–8.

Many people think that the 6-mil plastic often seen installed below a concrete floor is a waterproofing course, but its main purpose is to retard the set of the concrete (the concrete won't dry too rapidly). When any cement product (mortar, plaster, or concrete) dries too quickly, it's prone to shrinkage cracking. Without such a moisture-proof layer, the dry earth, sand in particular, will rob the moisture from the concrete so rapidly that it will be difficult to apply a good trowelled finish before the concrete becomes unworkably stiff.

In northern climates at least, the home should have a thermal break to the soil below using 1" of extruded polystyrene beneath the pour. We didn't use the polystyrene at the Cave, but we should

Illus. 5–8. Perforated 4" tubing was laid to remove any water that might accumulate under the floor.

have. The floor could have been kept a few degrees warmer in the winter. Happily, extruded polystyrene (Styrofoam®) is *close-celled,* which means that water won't pass through it. Styrofoam® can thus be used as flotation material for rafts or floating docks. Expanded foam (*beadboard*) isn't recommended in horizontal applications underground, because such foam can become saturated.

Because of the impermeability of the extruded foam, it's not necessary to use the 6-mil plastic as well. In the southern U.S., if heat dissipation is more important for comfort than heat retention, eliminate the foam from under the floor, or confine its use to within 48" of the footing. In either case, in order to retard the concrete set, use 6-mil polyethylene over any exposed sand. Don't worry about a few punctures in the plastic, as it really isn't meant to function as waterproofing.

UNDER-STOVE VENTS

Wood stoves need air, or they'll choke from lack of oxygen. We laid 4" nonperforated flexible tubing under our floor and took it up through the floor under the planned location of the two woodstoves. These two tubes connect with each other by a tee and travel under the footing to the lower of the two retaining walls near the door. The inlet is covered with ¼" hardware cloth. Supplying air directly to the stoves from the outside prevents drafts and keeps the stoves from robbing oxygen from the interior air. Also, with the tube buried underground for 20' or more, the earth actually preheats the combustion air quite a bit on its way in.

Try to avoid a "U" trap in this 4" intake line as it goes under the footing. Condensation can eventually fill the trap and stop the airflow. If you can't avoid the "U," it might be better to use perforated pipe so that any condensate will leach into the sand. We occasionally had to "bail out" the U-trap we'd erroneously created under the south-wall footing by using a rope to snake a towel back and forth through the pipe. Avoid our mistake, but do provide direct air to wood stoves.

UNDER-FLOOR PLUMBING

Now's the time, the only time, to install the waste plumbing. Ours was a simple system. The main waste line to the septic tank is a 4″ Schedule-40 ABS plastic pipe. In the house, it's common practice to tie all the exhaust lines from fixtures to a 3″ line (called the *house drain*) which expands to a 4″ line outside the house. The toilet flushes directly into the 3″ pipe. Other fixtures, such as bathtubs and washbasins, should have 1½″ outlet pipes, which join the 3″ pipe with a Y-connector made for the purpose. A vent stack, necessary for proper operation of the drainage system, should be a 2″ pipe. In large houses with lots of fixtures, there may be a main vent stack of 3″ or more going up through the roof. Check your plumbing design with the local building-code enforcement officer. The vent's purpose is that of a pressure equalizer. Without the vent, sinks may drain slowly and toilets won't flush properly. Odors can also find their way back to the house from the septic tank.

We installed a clean-out just in front of the toilet and hid the clean-out beneath the floor (Illus. 5–9), enabling us to run snakes or ramrods straight to the septic tank from the bathroom, should that ever be necessary. One of Murphy's laws states that if you don't put it in, you'll wish you had. If you do install the clean-out, you'll probably never need it. An improvement, according to plumbing manuals and standard practice, would be for the vent stack to be on the septic-tank side of the toilet.

All waste plumbing must be set firmly in the compacted sand, so that no settling can occur. The proper slope for all underfloor pipes in the waste system is ¼″ per foot. This translates to a 1″ drop over the length of a four-foot level, or ½″ on a two-foot level.

For ease of connecting the toilet later on, the toilet receptacle should be placed exactly at fin-ished floor level. Be sure to use the *roughing-in distance* (from the wall) for the actual toilet you plan to use. Leave all other waste pipes sticking up a good foot above floor level and wire them to iron pipe or rebar to make sure they won't be upset during the pour. Cover exposed openings to prevent sand or concrete from getting into the waste-plumbing system. Most of the waste plumbing at the Cave is in the bathroom and consists of a bathtub drain, a washbasin drain, a vent stack, a toilet receptacle, and a clean-out. There are all sorts of different connectors, elbows, expanders, and clean-outs on the market, but for ease of installation by the owner-builder, I recommend sticking with ABS or PVC Schedule-40 pipe. These plastic pipes are tough stuff, they'll probably outlast cast iron, and they're easy to cut and glue. Be sure to use the correct glue for the pipe you choose. Find a hardware store or plumbing-supply shop with a good stock of parts and a knowledgeable clerk.

The only other waste plumbing at Log End Cave is the kitchen-sink outflow pipe. For convenience, and to lessen the load on the septic system, we ran this pipe under the footing and into the same soak-away that absorbs water from the under-floor drains. Check to see if such a layout is acceptable in your area by contacting the town, county or state health department, whichever has jurisdiction. This simple grey-water system shouldn't be used with garbage disposals.

Only waste plumbing is discussed here, as its installation must be accomplished before the floor is poured. Try to consolidate your waste plumbing. If possible, the bathroom and kitchen plumbing should be on opposite sides of a common wall. If you're fearful about all these connections being underneath 4″ of reinforced concrete, then that portion of the floor containing the waste plumbing could be boxed in and a wooden floor installed later, so that access can be gained by using a screw gun instead of a jackhammer. On all the houses I've built, I've taken the time to install the waste plumbing carefully, compacting around each pipe, and testing the system before pouring the concrete. So far, no problems.

Supply plumbing is generally done *after* the floor is poured. Before the floor pour, provide a pipe into the home from your water source. A 1½″

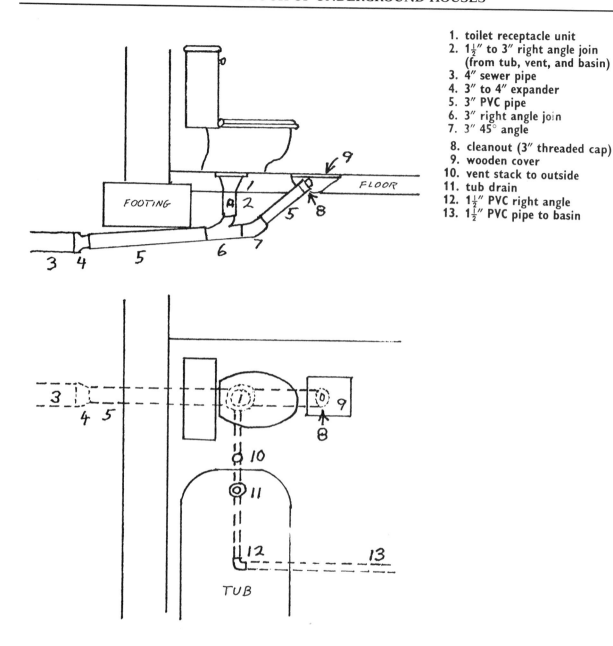

1. toilet receptacle unit
2. 1½" to 3" right angle join (from tub, vent, and basin)
3. 4" sewer pipe
4. 3" to 4" expander
5. 3" PVC pipe
6. 3" right angle join
7. 3" 45° angle

8. cleanout (3" threaded cap)
9. wooden cover
10. vent stack to outside
11. tub drain
12. 1½" PVC right angle
13. 1½" PVC pipe to basin

Illus. 5–9. The under-floor plumbing

drinking-water-quality flexible pipe should be more than adequate.

After the plumbing, under-floor drains, and stove-air inlets are installed, the 1" rigid-foam insulation can be placed. To minimize energy "nosebleed," try to keep the foam sheets joined together as tight as possible. Use 4' × 8' sheets, easier to use (and with fewer joins) than the more common

2' × 8' sheets. Don't put insulation over the half-poured pillar footings. Look at Illus. 5–7 again. The footings should already have Blueboard® beneath them.

WIRE-MESH REINFORCING

The purpose of the wire mesh is to hold the concrete together when it cracks. Most large floor pours crack despite all the best efforts to prevent it. Cracking is normal. We've had minor cracks in almost every floor we've poured, and so does everyone else.

The mesh is called 6 × 6 × 10 × 10 reinforcing mesh. This means that the matrix is composed of 6" × 6" squares, and the wire size is #10 in each direction. You can buy the mesh in rolls or in flat sheets, usually 5' × 10'. The flat sheets are easier to use. At the Cave, we actually used "goat fencing" of similar dimensions. Hundreds of feet of the fencing ran through the woods on our property. Someone many years ago had made a corral in the woods. All we had to do was cut the wire and drag it out. Recycling cuts costs. This reinforcing can be seen in Illus. 5–10. The regular stuff, used at Earthwood, can be seen on top of the insulation layer in Illus. 5–7.

THE FENCE

All that remains to prepare for the pour is to build a *fence* down the middle of the floor area to support the screeding planks. The fence cuts the narrow dimension of the pour in half, making screeding possible. Drive 2 × 4 stakes firmly into the ground and nail a row of 2 × 4s to these stakes so that the top of the horizontal 2 × 4s is at the same level as the top of the first course of blocks. Cut off

Illus. 5–10. Wire mesh holds the floor together if for any reason cracks develop in the concrete.

any stakes sticking up above the top edge of the fence. Prepare a couple of screed planks by notching out the appropriate amount of a 2 × 10, 6″ in our case. We used the same 2 × 10s that were used as footing forms. The fence and screed boards are seen in Illustration 5–11 and 5–12. With the contractor's level, check the fence for elevation.

CONCRETE CALCULATIONS

Finally, calculate the volume of the floor. With a 4″ floor the cubic yardage will be one-ninth of the square yardage (4″/36″ = ⅑). In our case, I knew by taking several test measurements that our floor would average close to 5″ in depth, and figured accordingly: 5″/36″ × 101 square yards = 14 cubic yards. I added two cubic yards for the deep bottom portion of our eight pillar footings, giving 16 cubic yards. We knew the job would require two loads, so we ordered 9 yards for the first load, in case our estimates were a little off. Nine yards did a little more than half the floor. Eight yards on the second load finished the job almost perfectly. Again, there was about a wheelbarrow-load left over. So we actually used about 16¾ cubic yards instead of the

calculated 16—about 5% more. I think it's reasonable to allow about 5% extra concrete in case of measuring (depth) error. And remember: A 4″ floor is plenty. If you've got an honest 3″ everywhere, you've got a good floor. Use a depth of 5½″ if in-slab heating pipes are to be installed.

POURING THE FLOOR

I hope you don't experience the last-minute rushing around that we had to go through. We were up at 5:00 A.M. finishing off the plumbing, and we finished levelling the fence about noon, while the cement truck was mixing on site.

Get the same helpers you press-ganged for the footing pour, and tell them you've really got a job for them this time. Experienced concrete pushers are hard to find. Again, for a strong floor, ask for a stiff mix. With the tines of a rake, draw the wire mesh up into the concrete as you pour (Illus. 5–11). Keep the mesh within an inch or two of the Styrofoam® to provide good strength and lessen the danger of getting the power-trowel blades caught up in the mesh later on. I always employ a designated wire-puller-upper, to make sure that the job

Illus. 5–11. It should be someone's job to make sure that the wire mesh is pulled up into the concrete.

gets done. Mesh sitting on the bottom of the pour won't do the job it's supposed to.

Do one side of the fence at a time. As soon as you've got a fair area poured, say, 20 percent of one side, start screeding. We used a power screed, which is nothing more than a big vibrator to which two wooden planks are attached. Using the power screed was only moderately successful. I think our planks were too long. To draw the screed along required another man at each end of the planks helping to slide them along the fence and the first course of blocks. The power screed might have been better if used on a shorter span or with smoother supporting runs (not blocks), but we found out later that two men with a single plank could do as well as the power tool (Illus. 5–12), saving a man's labor (and the cost of renting the power screed, if we'd known).

Water can be drawn to the surface and the floor smoothed further with a "bull-float," also shown in Illus. 5–12. A bull-float is a large, perfectly flat masonry float, usually made of aluminum or magnesium. Extension handles can be added to avoid walking on the wet concrete. I borrowed a bull-float, but they can be rented inexpensively from a ·tool-rental store.

After the first pour, there was a good hour's wait for the truck to make its return. We used the time to catch up on screeding and to have lunch. After a while, though, the welcome rest became worrying. The first pour was setting rapidly and there was still no sign of the second load. We covered the first pour with Styrofoam® sheets in an attempt to retard the setting, watered the edge where the two pours would join, and inserted 16" iron reinforcing rods 8" into the edge of the first pour to help knit the two pours together. At last the truck arrived and we were back at work, refreshed by the break. Still it was a little scary; if the second load had been delayed another half-hour, it might have affected the quality of the join. The moral: Specify two trucks. Usually, the trucks are radio-equipped, but if they're not, use the nearest phone to shout for the second load as soon as you can make a good estimate of what's needed to finish the job. Using Styrofoam® or plastic beneath the floor (as we did at Earthwood), you won't have this problem, since the water won't be drawn into the sand as it was the day we poured the Cave floor.

It was a long day. Drawing some 17 yards of concrete with rakes, screeding it, battling with

Illus. 5–12. A power screed can be used on a short span of concrete, but in this case two men screeding a single plank did just as well.

wire mesh, working around plumbing and under-stove vents . . . it's tough. And you, the owner-builder, are responsible for the success or failure of the project. You must coordinate the various jobs and make the decisions. I was extremely for-tunate to have Jonathan Cross on hand to keep us right. His experience and take-charge attitude probably made the difference between a so-so floor and the excellent floor we ended up with. If you can enlist the help of someone who has experi-ence with concrete slabs, jump at the chance. An-other time, someone might ask you for the benefit of your experience.

By five o'clock, we were ahead of the work and everyone went home except Cross. We had supper and took it easy for an hour, Jonathan checking the set every few minutes to see if we could commence power-trowelling. He impressed upon me the im-portance of experience in the handling of the power trowel, a machine which can run away with the operator if one of its four rotating blades digs into the concrete. I was thankful that he was will-ing to run the machine for us. I sprinkled water or cement as needed, and worked the edges with a hand trowel. One trick in obtaining a smooth sur-face is to scatter a dusting of dry Portland cement on the floor. The power trowel then does an excel-lent job of producing a flat, smooth floor, and I highly recommend its use.

We finished after 10:00 P.M., illuminated by car headlights. The next day was Sunday, and we made sure it was a day of rest.

6

External Walls

WALL CHOICES

The external walls can be made of stone, reinforced concrete, or concrete blocks. Building with stone is labor-intensive. Three families in our community built stone basements at about the same time we built our underground house. All three homes were slightly smaller than the Cave. The man with masonry experience built with a wooden form on the outside and free-form masonry on the inside. Working like a man possessed, he completed his walls in three weeks. The other two families spent the entire summer on their walls, and I doubt if either of the basements would meet waterproofing requirements for finished underground living space.

As much as I like stone masonry, I have some doubt about the economy of building with stone as opposed to blocks. One of the couples, using the slip-form method popularized by Helen and Scott Nearing (authors of *Living the Good Life*), used over two hundred bags of cement. They hauled in sand from a local pit for free and had old stone walls near their site. Using stone masonry is fractionally cheaper than using concrete blocks only if a zero dollar-value is placed on labor. Finally, the outer surface needs to be *parged* (coated) with a smooth layer of cement plaster prior to application of the waterproofing membrane.

Poured concrete is the common material used by most of the well-known advocates of underground housing. They're good architects, for the most part, but they don't actually perform the physical labor. For the owner-builder, my view is that the extraordinary forming involved with poured concrete is hardly worth the gain in structural strength, particularly since the methods described in this book are already stronger than what's needed to bear the loads and stresses in-

volved in underground-home construction. I'm reluctant to advise inexperienced builders to form and pour their own walls. The stresses on the forms are incredible. One owner-builder of an underground house near us poured his own walls. During the pour, there was a "blow-out" of the forms, and several cubic yards of concrete ruptured onto the floor. This sort of disaster could spoil your whole day!

We got bids from three contractors to pour the Cave walls, the low bid being $3000 in 1977. Triple that today. Our 12" block walls as described herein cost us $911 in materials, including surface-bonding cement, and $150 for hired help. While poured concrete may be stronger than surface-bonded blocks, the extra strength doesn't justify the much higher cost. We're very happy with surface-bonded walls, and we used the technique again at Earthwood. Finally, a poured wall takes quite a while to fully cure, adding to the home's humidity for a year or two.

People ask me about using cordwood masonry for below-grade walls. I'm not enthusiastic. The secret to cordwood's longevity is its ability to "breathe" along the end grain. When one side of the cordwood is waterproofed, this breathing ability is lost. Also, cordwood is weak on tension. Its resistance against lateral pressure, then, is low. A curved wall addresses this problem, but the issue of lack of breathing still remains. Finally, cordwood-masonry walls, while beautiful, are light-absorbing. A bright white wall maximizes light in places where natural light is minimal. Malcolm Wells says:

One of the easiest ways to stop wasting energy comes in buckets. It's called white paint, which, when applied to darker walls, makes rooms so bright you need only half the light bulbs you formerly used.[7]

This leaves blocks. A conventionally mortared and a surface-bonded wall cost about the same when all factors are considered. The mortared-block wall requires more skill and more labor than does the surface-bonded wall, yet the surface-bonded wall is much stronger than is the mortared wall against lateral pressures. For a home with straight walls, use 12″ concrete blocks in conjunction with surface bonding. This method is strong, moderate in cost, easily learnable, and relatively fast, and it supplies an excellent base for waterproofing outside and painting inside. These are the reasons I accent surface-bonded block walls here.

BUILDING SURFACE-BONDED BLOCK WALLS

My nephew Steve Roy and his friend Bruce Mayer arrived at Log End during the wall construction. They were to become indispensable additions to the work team during the next three weeks.

With the first course in place and level, succeeding courses are dry-stacked; that is, they're laid

Illus. 6–1. Before stacking, a piece of broken block can be slid across other blocks to knock off excess materials and burrs.

without mortar, block tight against block. Each block should cover the join between blocks of the previous course. The tops and bottoms of blocks should be cleaned of burrs by rubbing blocks together, or by rasping them with a piece of broken block (Illus. 6–1).

Stack the blocks three courses high at the corners (Illus. 6–2), and check for plumb using the plumb bubble of a 4′ level. Then fill in between corners with blocks (Illus. 6–3), always working to a mason's line stretched tight between corners. Repeat the procedure with another three courses of blocks, and so on.

If a little wobble is perceived in a block, shim one of the corners with a thin metal shim cut from a piece of aluminum offset-printing plate or aluminum flashing. Use your level to determine which of two corners is best to shim. Sometimes a doubled or folded shim may be necessary.

Buy ready-mixed surface-bonding cement, available from several different manufacturers (Conproco, W. R. Bonsal Co., Quikrete, Stone Mountain Manufacturing Co.). Ask at your local masonry-supply yard for *surface-bonding cement*. They may call it *block bond*, a holdover from the days when there was a product by that name.

On a single-storey home like Log End Cave, you can actually stack all of the blocks before any surface-bonding cement is applied. With 12″ blocks, you'll be impressed at how strong and stable the wall is, even in its dry-stacked state.

Soak the wall with water (Illus. 6–4) before you mix the bonding cement. This soaking prevents overly rapid drying of the surface-bonding cement. The wall should be saturated, but not dripping. Now mix the cement according to the manufacturer's instructions. The cement should be quite wet, like very thick paint, but not so wet that it falls off the wall during application.

Apply the surface-bonding cement to the wall using a flat trowel as shown in Illus. 6–5. Load the flat trowel using a pointed mason's trowel. Apply the mix with firm trowel pressure, pushing the load upward and outward until a uniform coverage is attained. To even out the plastered area and to spread excess cement, follow with longer, lighter strokes, holding the trowel at about a 5° angle to the wall surface. Don't over-trowel, as this can cause cracking or crazing. It takes a little practice, but the technique can be learned fairly quickly.

Illustrations 6–2 and 6–3. After the first course of blocks is in place and level, build the surface-bonded wall by stacking the blocks three courses high at each corner (above); then fill in between the corners (below).

Illus. 6–4. Before applying the surface bonding, spray the wall with water until it's wet, but not dripping.

Illus. 6–5. Work the mix onto the wall using a plasterer's trowel.

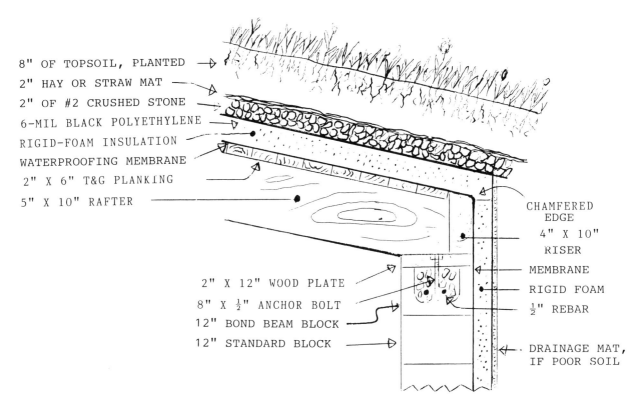

8" OF TOPSOIL, PLANTED
2" HAY OR STRAW MAT
2" OF #2 CRUSHED STONE
6-MIL BLACK POLYETHYLENE
RIGID-FOAM INSULATION
WATERPROOFING MEMBRANE
2" X 6" T&G PLANKING
5" X 10" RAFTER

CHAMFERED EDGE
4" X 10" RISER
MEMBRANE
RIGID FOAM
½" REBAR
DRAINAGE MAT, IF POOR SOIL

2" X 12" WOOD PLATE
8" X ½" ANCHOR BOLT
12" BOND BEAM BLOCK
12" STANDARD BLOCK

Illus. 6–6. Detail of the top of the fully bermed walls, 40' × 40' Log End Cave plan

The inner surface of a rectilinear underground building should have an average surface-bond coating that's ⅛″ thick, which depth can be checked with the corner of your trowel (until you get the knack). The side of the wall facing the earth need only have an honest 1⁄16″ coating of cement, as the backfilled side of the wall isn't in tension. The wall can only collapse inward, not outward. Above grade, use a ⅛″ coating on *each* side of the wall, because you can't be sure in which direction failure might occur.

After application, mist the surface-bonding cement frequently, at least twice within the first 24 hours, more frequently if the walls are in the sun. Keeping the wall wet will greatly reduce the chance of shrinkage cracking or crazing. Try to work in the shade, if you can, paying attention to the sun's location at different times of the day.

OPTIONAL BOND-BEAM COURSE

A good way to impart extra tensile strength to a long, straight earth-sheltered wall is to incorporate a *bond beam* on the final course (the 11th course on the 40 × 40 plans). The bond beam is made by laying out a course of special bond-beam blocks, shown in Illus. 6–6 (an important drawing, containing much information. This illustration will be referred to again during discussion of roofing, waterproofing, and insulating). Bond-beam blocks have a U-shaped cross-section. The trough formed by several consecutive blocks is filled with concrete and ½″ rebar, as shown. This solid-concrete bond beam is very strong in resisting lateral pressure. Even more tensile strength can be gained by incorporating a bond beam on another course about halfway up the wall.

Concrete for filling the bond-beam course, or for filling block cores where the vertical rebar is to be installed, can be made in a wheelbarrow, using either a premixed concrete mix such as Sakrete®, in bags, or by mixing your own concrete, using a recipe of 3 parts #1 stone, 2 parts sand, and 1 part Portland cement.

SURFACE BONDING AS WATERPROOFING

While surface-bonding cement has been rated excellent for resistance to wind-driven rain at velocities up to 100 m.p.h., and has withstood a 4' hydrostatic head with no water penetration, it shouldn't be considered to be an absolutely waterproof membrane for underground housing. The problem with any cement waterproofing is that its integrity depends upon there being no break in the crystalline structure of the waterproofing material. If there are any crazing, settling, or shrinkage cracks, the membrane is broken. However, surface-bonding cement certainly provides an excellent base for the application of a sheet membrane, and may have some value as a second line of defense under such a membrane.

PLATES

The changeover from masonry to timber construction is accomplished by the use of wide wooden plates (*sills*) bolted to the top of the masonry wall. With a stone wall, the anchor bolts must be built in with the last course of stones. With a poured wall, the bolts are set into the concrete while it's still workable. There are two ways of setting the anchor bolts into the blocks' cavities, and we employed both, depending on the location of the bolts. In corners and other locations where the cavities were completely filled, the bolts are set in the concrete while it's still workable. If an anchor bolt is required between filled-core locations, stuff a wad of crumpled newspaper 18" into the hole and fill it with concrete. If a bond-beam course of blocks is used to finish off the top of the wall, the anchor bolts can be set right in the concrete-filled

trough and hooked under the rebar, as shown in Illus. 6–6. Use a 12" anchor bolt having one threaded end and an ell bend at its opposite end. The ell assures that the bolt can't be pulled out once the concrete is set. Leave the bolt exposed above the wall a distance equal to the thickness of the plate to be used.

One way to make sure that the bolts are plumb and at the right height is to make a template out of a 12" piece of the same material from which the plates are made, 2 × 12s, for example. Drill a hole the same diameter as the bolt, or slightly larger. Take care that the hole is perpendicular to the top of the plate. This template can now be used to check both depth and plumb.

Plan the location of the anchor bolts based on the lengths of the planks you'll use for plates. Two bolts are sufficient for short plates; use three for plates over 9' long. Keep the bolts about 12" in from each end of the plate.

I like wooden plates that are at least 2" thick. Rough-cut 2 × 12s are perfect for the Cave plan being discussed. Ordinary dimensional 2 × 12s are actually only 11¼" × 1½", which makes them just a little narrow at the top of a 12" block wall. At the Cave, we used 4 × 8 plates on the east and west walls, to accommodate the 4 × 8 rafters (*roof joists*). The 40' × 40' plans, however, call for 5" × 10" joists. Moreover, I now place all rigid-foam insulation on the exterior of the roof- and wall-waterproofing membranes, so another adjustment has been made from what we actually did at Log End Cave. The new detailing is shown in Illus. 6–6.

After the anchor bolts are permanently set in the concrete (allow two days), locate the position of the holes in the plate by laying the plate on the anchor bolts and whacking once sharply with a heavy hammer (Illus. 6–7). Have a helper hold the plate steady and in the correct position while you do this. The indentations formed mark the location of the holes to be drilled. Turn the plate upside down and drill holes of the same diameter as the bolts, being careful to keep the brace and bit (or drill) straight up and down (Illus. 6–8). Next, chisel out a depression to accommodate a washer and a nut (Illustrations 6–9 and 6–10). Then lay a thin layer of flexible polyethylene sill-sealer on the top of the wall where the plates will go, and bolt the plates down with a crescent- or socket wrench,

Illustrations 6–7, 6–8, 6–9, and 6–10. After the anchor bolts have set in the concrete, locate the holes in the plate by laying the plate on the anchor bolts and whacking once sharply with a sledgehammer. The indentations in the plate mark the locations of the holes to be drilled. Turn the plate upside down and drill holes the same diameter as the bolts. Chisel out a depression to accommodate a washer and nut.

compressing the "sill seal." The sill seal, which you can buy at building-supply yards, protects against infiltration by drafts and insects.

To finish the wall, I recommend the installation of a riser piece, on edge, as shown in Illus. 6–6. This riser piece will be made of two or three lengths of 4" × 10" timbers butted up to each other to match the length of the 30' or 40' sidewalls. The riser pieces should be somewhat dry, so as to not trap a lot of moisture after waterproofing, and they can be "toe screwed" to the top plate, or fastened with angle-iron braces. In this detail, the advantages of using a riser piece, as opposed to

mortaring blocks as we did at the original Cave, are that the inside looks better (you don't see bare concrete blocks) and installing a riser piece is much quicker than mortaring blocks.

The only drawback is that the wooden riser pieces might be more expensive than working out a detail using blocks. If time is money, though, the little extra cost, if any, is more than offset by the advantages. Note the chamfered edge at the top of the riser piece, so that the waterproofing membrane can more easily bend around this corner. The planking on top of the 5" × 10" rafters comes to the same level as the top of the riser.

7
Timber Framework

POST-AND-BEAM

Post-and-beam construction is my favorite framing technique, not only because it has the strength to support the kinds of loads dealt with in underground housing, but because one can *feel* the strength in an almost ethereal way, particularly in structures where the framing is exposed. The same holds true for plank-and-beam roofing, the natural companion to post-and-beam framing. Most people like the rustic atmosphere of old inns, with their exposed hand-hewn timbers enclosing plaster panels. I think we pick up a sense of security from seeing those massive beams overhead. We know that they aren't going to fall in on us.

OLD BARN BEAMS

You can have timbers milled from selected logs, or you can use old hand-hewn barn timbers. I chose barn timbers for the post-and-beam framework and had the roof rafters and 2 × 6 roof decking milled from straight hemlock trees which I selected myself. There were three reasons I decided on barn timbers:

- They're incredibly strong. Most of the main support beams during the 19th century were hewn to 8″ × 8″ or 10″ × 10″ dimensions, and the trees were finer-grained in the old days, before planned forestation was introduced to encourage rapid growth. Of course, any beams that show rot, whether from insects or moisture, should be rejected immediately. Choose carefully, and store all wood up off the ground while it's waiting to be used. With barn beams, stack them so that water cannot collect in the hollows

(*mortises*) that were chiselled out for mortise-and-tenon joints.

- Old barn timbers are no longer green and they've finished shrinking. Using dry wood may not make a lot of difference with some kinds of construction, but our intention was to use a lot of cordwood masonry to fill many of the panels framed by the posts and beams. Green wood will shrink away from the masonry infilling. The wood masonry units used with this style of construction are called *log ends*; thus "Log End Cave."

- I like the character of the old beams: the mortises, the adze marks, the dowel holes. The wood's character is imparted into the atmosphere of the finished room.

Which old barn beams to use is a value judgment, and I'm fastidious in their selection. We find plenty of nonstructural uses around the home for "rejects." Our strategy, which we've used at three of the houses that Jaki and I have built, is to carefully examine and catalog each beam, noting its dimensions and quality. Using this list as a guide, the timber framework is then designed on paper. If we're a beam or two short, we seek out replacements; we don't try to make a questionable timber serve a structural purpose.

ROUGH-CUT TIMBERS

If you're building in an area with building codes, the use of recycled timbers for structure may not be allowed by the code-enforcement officer. New rough-cut posts and beams from a local sawmill are very strong, economical, and can be quite attractive, especially if bleached after a year or two in the building. Jaki bleached many of the rafters and floor joists at Earthwood using a toothbrush (!) and a 50% bleach-and-water solution. Some building inspectors may require that wood from a sawmill be graded (inspected and evaluated). Find out before you start.

We looked for suitable rafters among old barns and houses that were being torn down in our area, and found a few rafters that would have done the job, but not enough of them were of consistent size, so we decided to use new 4″ × 8″ hemlock rafters. Shrinkage on exposed rafters doesn't really affect the construction in any way, and, as we were able to choose our own straight trees, we knew the timbers would be strong. We also bought the wood for the planking from the same supplier, a local sawmill.

A year or two later, I learned that hemlock is one of the weakest woods in terms of shear strength. Shear doesn't normally come into play with light-frame construction, but with heavy loads it does. Our hemlock's low shear strength was the weak link in the entire structural system at Log End Cave. In fact, calculations later revealed that the rafter shear strength was technically less than that required. The building never seemed to suffer any for this, and I credit safety margins in both formulae and in wood grading as the reason for our having no problem. The new 40′ × 40′ plans are engineered correctly for both shear and bending for the spans and loads required. Have *your* plans checked by a structural engineer.

Green wood should be stacked on site (with wooden laths or *stickers* between courses) as long as possible before using it. Although shrinkage on the rafters doesn't matter much (they don't shrink at all on length), you don't want ½″ spaces between the roofing planks. If I were doing the Cave again, I'd use seasoned spruce or pine tongue-and-groove 2 × 6s. In northern New York, they cost $730 per thousand board feet in January 1994, which works out to be about $1460 per thousand square feet. With two-by material, one square foot contains two board feet.

If you're lucky enough to find an old silo that's being pulled down, as we did when we built Log End Cottage, you'll have good roof planking: old, dry, full-size, tongue-and-groove 2 × 6 planks. Spruce, if you're really lucky. Ask around. Some farmer may be willing to sell a leaning silo which he's not using anymore. We bought a silo for $100. Figure the board feet and make an offer. And don't forget: Those old silo hoops make great rebar for footings and block-wall cores.

A big asset to building economically is being a good scrounger. Old recycled timbers are often superior to what's being produced today. What's more, you'd be saving trees.

We built the Cave around the three major support members, 30′ 10 × 10 barn beams, which we'd bought 25 miles away for a dollar a running foot. They're beautiful timbers, exposed throughout the house, a real design feature of the construction. We tried to plan our post location to take into account both structural considerations and practicality with regard to the floor plan. We chose the beam with the biggest cross-section as the center-support girder. I didn't want to span more than 10′ with an old barn timber, although it was in excellent condition and was a full-size 10 × 10, so I planned for two interior support posts, one out of the way on the stone hearth, and one between the kitchen and dining areas. All three spans thus created are less than 10′. Nowadays, I actually perform the stress-load calculations before building, but I didn't know how to do that back in 1977.

INTEGRATING THE STRUCTURAL AND FLOOR PLANS

The floor plan was designed so that the north-south internal walls would fall under the other two 30′ beams, so they're each supported by three major internal posts, dividing the depth of the house into approximate quarters. We placed the posts at the intersections of the internal walls, so that the 4″-thick walls could butt against the 8″ × 8″ squared posts, leaving the posts exposed in the corners of the peripheral rooms. The two peripheral 30′ girders are positioned exactly between the center-support girder and the east and west sidewalls, so that the clear spans of the rafters are all consistently 8′. A big advantage of this plan is that the long north-south internal walls rise up to meet the underside of the big girders, so there's no tricky fitting of the walls around the exposed rafters. The exception is where Rohan's room jutted out into the living-room area, necessary to give him a reasonably sized space (Illus. 7–1).

Winter at Log End

(Left) Log End Cave faces due south, West Chazy, N.Y. (Left, bottom) The stoves and stone mass at the center of Log End Cave (Below) North gable end of Log End Cave, with a stone circle made from boulders from the excavation

A

The "library wall" with 14 different species of log, "Earthward," Minnesota (John Rylander photo)

The Log End Sauna was the author's first freestanding earth roof, 1979.

"Earthwood Junior," a single-storey version built by Paul Mikalauskas in Ashland, N.H. (Paul Mikalauskas photo)

Earthwood in winter

"Earthward" interior (John Rylander photo)

(Left, middle) Earthwood Building School, West Chazy, N.Y. (Left, bottom) Freestanding grass roofs on Earthwood outbuildings

The Underground Studio, Jay, N.Y.

The earth roof at the Underground Studio, Jay, N.Y., with a view of the Adirondack Mountains

Another view of the Underground Studio

D

Illus. 7–1. One room juts out into the living room, the only place where the walls had to be specially fitted around the exposed rafters.

All the east-west walls rise up to meet the underside of 4″ × 8″ rafters, except for the wall between the office and the master bedroom, which meets the ceiling halfway between two rafters. There are two good reasons for integrating the structural and floor plans: Integrating the floor plan with the support structure saves slightly on materials, because the 8″-high rafter is used as a part of the internal wall; and it results in something that looks better than just missing a rafter with a wall (which would also form a great space for cobwebs). Because of space considerations, we varied slightly from this plan with the office-bedroom wall, splitting the difference between rafters in this case.

SETTING UP THE POSTS

The building plan called for a 22″ space between the top of the block wall and the underside of the center girder. Subtracting 2″ for the plate, we were left with a 20″ post to support the north end of the girder (Illus. 7–2). We called the floor level and based the heights of the two internal support posts on the total height of the girder off the floor at the north end, 95″ (7′11″) to its underside. We cut the two interior freestanding posts to 96″ (8′), so that we could recess them an inch into the underside of the girder. I wouldn't do this again, as that 1″ check

into the girder decreases its shear strength by 10%. Just toenail the top of the post into the girder. Even easier than nailing is to drill a pilot hole on a 45° angle and install 4″ steel deck screws, using a cordless drill instead of toenailing.

In 1993, I finally purchased a hand-held rechargeable (and reversible) drill. It has become my most-used tool. I hardly nail anything anymore. It's easier and stronger to use screws, and screws can be removed easily. A great variety of screws is available today, made for every purpose, from hanging sheetrock to fastening deck boards.

Since the south-wall post was in two pieces, it was a bit trickier to figure. The base post, the one standing on the wooden plate, had to be 48½″ to accommodate the 48″-high windows we'd planned for the south wall. The header for the windows was to be a 4 × 10 laid on its side, the 4″ dimension to resist against sagging, 10″ to establish the width of the cordwood-masonry wall later on. The short post on the top, then, was all that was left of the 95″ of height established at the north end. The subtraction was: 95″ minus 28½″ (the short block wall) minus 2″ (the plate) minus 48½″ (the base post between windows) minus 4″ (the window header), leaving a short "post" of 12″. The calculations checked in the field. A mason's line stretched from the top of the north-wall post to the south-wall post showed one of the interior posts sticking up 1″, the other ⅞″. This was just right for our intended purpose of checking or keying the posts into the girder. The floor was level after all!

Having removed all nails and marked the rough-hewn posts as best we could with a square, we cut one end of the post with a chainsaw and stood the post upright to check for plumb. If the post tilted in any direction, it was necessary to recut the bottom, compensating for the error. Only when the bottom of the post was flat did we measure for the cut at the top. You can use the square again on the top measurement, but I've found that it's less risky to measure the length of the post from the good end along each of the four corner edges, and then connect the marks. The irregularities of a barn beam can throw the square off, and there's little room for error on the second cut. You won't have such problems if you use milled beams.

Don't stand wood directly on a concrete floor, because of the possibility of trapping dampness. Cut a square of 90-pound roll roofing or asphalt shingles to stand each post on. W. R. Grace's Bituthene® waterproofing membrane, described in the next chapter, is also excellent for this purpose. This damp-proof shim also helps to steady the post. To lessen the chance of someone being hurt by a falling timber, we waited until all eight internal posts were cut and squared before standing them up permanently. Calculating the heights of posts that support the two peripheral girders is best explained with the aid of Illus. 7–2.

location is established on the plate such that B = B'. Now consider the total height of the supports for the 4 × 8 as measured from the top of the block wall. At the lower end, the rafter is to be supported directly by the 4"-high plate. The high end of the rafter is supported by a 2" plate, a 20" post, and a 10" beam—32" in all. The height of X, then, should be midway between the 4" and the 32" figures, or 18". Subtracting 2" for the plate and 10" for the support beam leaves a 6" post. I cannot recommend actually using a "post" as short as 6", because a heavy load might cause it to split open, just as a thin slab of wood is easier to split with an axe than is a thick one. Instead, I used two short pieces cut from a scrap of an old 3 × 10, stacked one upon the other. These two short pieces worked perfectly: They were exactly the right dimensions and were easy to fasten to the plate with 20-penny nails. A measurement from the floor established the height of the internal posts, as discussed previously for the center beam.

The support structure of our house is actually a very simple one and I hope that by examining in some detail the ways in which we met our problems, the reader will gain a sense of the principles involved, enabling him to solve the structural problems of a design that's somewhat different from the one he might wish to employ. The mathematics involved is never more than simple arithmetic and the most basic geometry. Keep a set of plans nearby, make notes on them based on actual

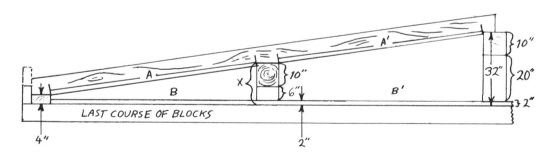

Illus. 7–2. Calculating the heights of the internal posts that will support the two peripheral beams

To make the best structural use of the 4 × 8 rafters, I wanted to make the spans A and A' equidistant. This will be accomplished if the post

(not theoretical) measurements, and refer to your notes constantly, double-checking your figures from two different approaches, if possible.

When all eight internal posts were squared and cut, we stood them up on their squares of roll roofing at the correct premeasured locations, shimming, if necessary, to assure perfect plumb. We made the posts rigid with a rough framework of 2 × 4s (Illus. 7–3). We didn't attempt to fasten the posts to the concrete floor. The tons of weight that these posts support, as well as the internal wall tie-ins, assure that the posts won't move once the roof is on. The rough framework's purpose is to hold the posts upright while putting the massive 10 × 10s in place. We didn't dismantle this temporary framework until after the rafters were fixed.

The window and door header (or lintel) on the south side of the house had to be put into place before the small post for the south end of the center beam could be permanently located. This lintel consists of a pair of 17-foot 4 × 10s, cut for the purpose at our local sawmill. It took four of us to maneuver the things into place. The south ends of the other 30-footers would rest directly on the headers at the points of their greatest strength: right over stout 8 × 8 barn posts (see south-wall elevational plan, Illus. 2–3).

The plank-and-beam roof of our house is carried by the east and west block walls and the three aforementioned 10 × 10s. The dimensions of the beams may vary quite a bit from beam to beam and from end to end, but the smallest dimension at any of the six ends is a full 8″ × 9″, and the average size is very close to a full 10 × 10.

The beams were stacked almost a quarter-mile from the Cave site. Steve, Bruce, and I took very careful measurements of the locations of the posts and transposed the measurements onto the actual timbers. We'd chosen the heaviest of the three beams as the center beam, and labelled the others "east" and "west."

We squared the ends of the beams. The total length of the center beam is 30′3″, which allows the ends to extend 4″ beyond the support posts (for the sake of appearance). The west beam was cut at 29′11″ to give a 2″ overhang at each end. We wanted to make the east beam the same, but we could only get a good 29′8″ out of it. We solved the problem later by adding a 3″ piece from another beam, which in no way affected the structural considerations, because the last rafter still had good solid wood to rest upon. This cosmetic surgery is only noticeable if it's pointed out, as all the external

Illus. 7–3. After the internal posts are squared and cut, stand them up on their squares of roll roofing at the premeasured locations and shim, if necessary, to assure perfect plumb. The internal posts can be kept in place with a rough framework of 2 × 4s.

surfaces of the beams were later creosoted to a uniform color.

I went out to recruit some able bodies while the boys finished up the checks in the beams to accommodate the internal posts. As usual, it was a hectic day of running around making last-minute preparations. By late afternoon, I'd gathered four strong bodies to supplement Steve and Bruce. We backed the pickup to the western beam, the lightest of the three. We figured that if we couldn't move that one, we'd have no chance with the 600-pound center beam. With Jaki looking after the baby, I was left with the easy job: driving the truck. Six men lifted the north end of the timber while I backed under.

The beam was now tucked right up to the back of the cab. Then, paired at each end of strong ash branches, the three teams lifted the south end off the ground and shouted the command "Go!" Like some great crawling insect, we started up the hill. Once, halfway up the hill, I was chastised by six screaming banshees, in no uncertain terms, to quit speeding. We purposely overshot the driveway and backed up to the house site. We backed the beam so that almost half of it was cantilevered over the edge of the north wall, and then manhandled it from below into position on the posts (Illus. 7–4). To our great delight, the post fit perfectly into the checks. So much for the easy one. The boys voted to tackle the biggie next, reasoning not unwisely that they mightn't be able to manage it after hauling another of the small 500-pounders.

Not only was the center beam considerably heavier than the other two, but it had to be raised almost 2′ higher. We manhandled it halfway out over the hole, as before, and then put our strongest and tallest man, Ron Light, below the beam, catwalking with Bruce along some makeshift scaffolding. Almost single-handedly, Ron passed the south end up to where Paul and I could get our ash branch under it. Then, with a monumental effort

by all, we hauled it up and into place. Again, the checks were beautifully positioned, but, alas, one wasn't deep enough. "It's gotta come off," I said. Groans from the crew. "Maybe the post will compress," suggested one of the tired backs hopefully. We took the beam off the posts and laid it to one side while I used a chainsaw to remove ⅜″ from the top of a post. Steve did a little chisel work on the check. Back went the beam, this time fitting almost perfectly. We'd actually overcompensated slightly and the beam didn't rest on the post by ⅛″. Good enough. We knew the beam would sag ⅛″ under the roof load and rest on the post, as intended.

The last beam, by comparison, was easy, especially since I was still driving the truck!

PLANK-AND-BEAM ROOFING

Our 4 × 8 rafters are the "beams" of plank-and-beam construction. Why 4 × 8s? A 4 × 8 is proof against twisting under a weight load. The base of the rafter is so wide that it sits unassisted on the support beams. Toenails are sufficient to keep it in place. With two-bys, it's necessary to cross-brace adjacent rafters with metal ties or wooden blocks

Illus. 7–4. The beam was then cantilevered over the north wall and positioned on the posts.

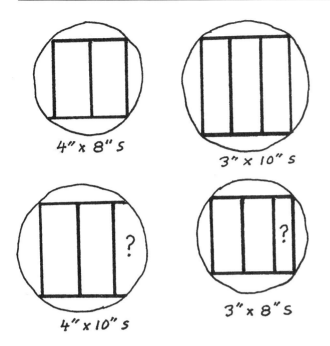

4" x 8" S

3" x 10" S

?

?

4" x 10" S

3" x 8" S

HEAVY LINES INDICATE SAWCUTS

Illus. 7–5. The advantage of 4 × 8 or 3 × 10 rafters: 3 × 8s and 4 × 10s don't return the maximum number of rafters from round logs.

to protect against twisting under a weight load. The 8″ measurement provides excellent strength on the relatively short 8′ spans and has the advantage of being twice the 4″ measurement. This makes for economic use of the log and gives a well-proportioned rafter, pleasing to the eye if left exposed. The alternative size which we considered was a 3 × 10 rafter, as we'd used at the Cottage. These rafters, too, fit nicely into the cross section of a log, but they require larger trees, which may be difficult to find nowadays. The reader will readily see in Illus. 7–5 that measurements such as 3 × 8 and 4 × 10 don't return the maximum number of rafters from round logs.

In the new 40′ × 40′ Cave plans, the rooms are bigger than they were in the original Cave. Rafters span 10′ instead of 8′. These longer rafters give a lot more room in the bedrooms and bathroom, so 5″ × 10″ rafters are required instead of 4″ × 8″. But, because single spans are used instead of double spans, the actual timbers are shorter, lighter, and easier to handle. Single-span girders are now

recommended with the 40′ × 40′ plans, greatly diminishing the weight of the timbers that need to be manhandled.

We put our rafters 32″ *on center* (32″ o.c.), mainly for strength. "On center" refers to the distance from the centerline of one rafter to the centerline of an adjacent rafter. We tailored our floor plan and the east- and west-wall lengths to multiples of the 32″ centers. Putting rafters 32″ o.c. also worked in well with the blocks. Someone accustomed to stick framing and plywood may think that 32″ centers are rather wide and, for that type of building, they would be. But plank-and-beam construction permits up to eight *foot* centers (96″ o.c.) to carry a "normal" roof. An earth roof, of course, isn't "normal," so we went with 32″ centers. The planking itself would be quite happy on 48″ centers, as can be seen at Earthwood, but there wouldn't be enough rafters to carry the heavy earth-roof load.

INSTALLING THE RAFTERS

The three girders were in place, but the rafters were still in the form of logs. At least they were at the sawmill. To speed up progress and get the wood out, the boys and I had to spend two days at the property where I bought the hemlock. The order was already seven weeks overdue. Bad enough that this had delayed the job, but it also meant that we'd be putting the plank-and-beam roof up green, causing considerable shrinkage. Have these materials lined up well in advance, particularly if you're dealing with small-scale loggers and sawmills, who sometimes take a relaxed attitude towards things like delivery dates. Even two months' drying makes a huge difference in the moisture content of lumber.

At Log End Cave, we "bird's-mouthed" all the rafters so that they'd rest with stability on the east and west walls, and the girders. A "bird's-mouth" is the space left when a wedge of wood is taken out of the underside of the rafter so that rafter has a flat bearing surface where it rests on the support structure below. I don't advise bird's-mouthing anymore, because it can decrease shear strength of the rafter by 10% or more. Shear strength is a function of the cross-sectional area of the rafter,

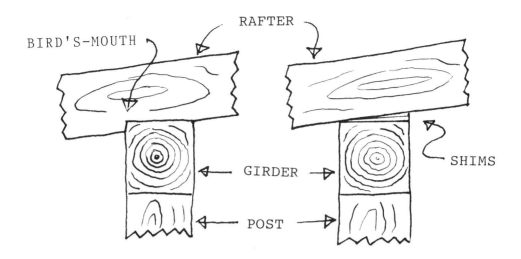

Illus. 7–6. Left: Cross-sectional area is lost, along with some shear strength, when a rafter is bird's-mouthed to rest solidly on a girder (or wall plate). Right: Shimming doesn't decrease the shear strength of the rafter.

and shear failure occurs not in the middle of a span, but right where it's supported from below by a wall or beam. So the bird's-mouth is at the worst possible place. At Earthwood, we shimmed the rafters with tapered cedar shingles instead of bird's-mouthing. Not only is shimming stronger than bird's-mouthing, it's very much easier to do.

Just place the rafter in position, and jam in the right amount of cedar roofing shingles to give good bearing (Illus. 7–6).

We used three men to slide the rafters into their positions. After snapping the picture (Illus. 7–7), I moved into position on the center girder so that the other men could slide the heavy rafter up to

Illus. 7–7. The rafter is slid along the beams and put into position.

Illus. 7–8. Half of the rafters are now in place. I waited until all were in place before fastening them.

me. Eighteen-foot green hemlock 4 × 8s are heavy, probably about 180 pounds each.

I waited until all the rafters were in place (Illus. 7–8) before fastening them. With rough-cut and barn timbers, the tops of the rafters are unlikely to be in the same plane for planking. Again, additional cedar shims on the "short" rafters can bring all the tops of the rafters into the same plane for planking. Then toenail or, better, "toescrew" the rafters in place.

We Interrupt This Story to . . .

If this were purely an instructional manual on how to build a particular house, it would be unforgivable to stop halfway through a discussion of plank-and-beam roofing for any reason whatsoever. But it's not simply an instructional manual; it's also the story of an owner-builder wrestling with a little-known style of building.

The completion of the rafters on August 26 marked the end of a productive period that began on July 5 when the digger took its first big bite out of the knoll. Except for the delay in getting the hemlock to the sawmill, progress had been swift, at least from an owner-builder's point of view. Jonathan Cross might have considered our pace to be that of a particularly sluggish snail. Still, fifty-three days after the start of construction, we were ready to put the roof on . . . if the roof had been ready for us.

I fell off a rafter onto the window header below, badly bruising, if not cracking, a rib. Fifteen minutes later, I was back on the beams, a wiser man. We finished the rafters, and I laid blocks the next day, but

four or five days after the tumble, I couldn't do any physical work, and the project came to a halt for a week. I used the time to try to solve the next problem, which was that we were about 15% short of the required roof planking. There was no way I was going back to my original supplier. I couldn't wait another six weeks for the wood. I was forced to buy five plantation pines and haul them to the sawmill. I decided that as long as I mixed the pine planks carefully with the hemlock, I wouldn't be sacrificing too much in the way of strength. I think this turned out to be a correct assumption, though the pine was very green indeed and began to go a little moldy in the interior until I connected the stove and dried it out. Today, the effect is that every fifth or sixth board has a kind of a bluish-grey color, rather nice, really, though not exactly what we'd planned. Over the sauna we used hemlock exclusively, fearing that the 200°F temperature might cause the pine pitch to bleed from the ceiling. Eastern hemlock isn't a pitchy wood.

The concrete blocks I laid were the first course of 4" blocks to go on top of the 12" block wall. The reader will recall that our east and west plates were only 8" wide in order to leave room for these blocks. I didn't use surface bonding because it could only have been applied to the wall's exterior. This portion of the wall wasn't loadbearing, so I felt confident in laying up the blocks with mortar. I decided to lay the first course before nailing the roof planking, so that the planking wouldn't be in the way of my work, but I held off on the second course so that the blocks wouldn't be in the way of the planking.

Here I should mention a problem the reader can easily avoid. We'd spread about half of the planks on the roof, loose, to give them a good place to dry. Before any rain, we laid out rolls of 15-pound felt paper, lapped, to try to keep as much water out of the house as possible. We couldn't start nailing the roof, as we were still waiting for the pine logs to be milled into planking. One night, the first strip of felt paper blew up and backfolded onto the roof. Water poured down the roof and onto the top of the 4" blocks, finding its way finally into the unfilled cores of the 12" blocks. There's no way to get that water out short of drilling holes into the block cavities (from the inside, so as not to break the waterproofing quality of the surface bonding). This we did. The moral: Maintain vigilance in keeping water out of the wall cavities.

Steve and Bruce left on August 28. Their help had been invaluable. I think my diary best captures the mood of the next twelve days:

"Monday, August 29. Hauled 2 × 6 planks from the sawmill." Progress was so slow that the next eleven days are all lumped together:

"Tuesday, August 30–Friday, September 9. Discover that we will run short of planking (thought we had more than enough), so I chase around for hem- lock. No luck, so I work a deal with Dennis for pine. What we think is Norway pine turns out to be planta- tion pine. Gary and Diane arrive on Labor Day week- end and Gary helps Dennis and me clear a road to the pine. I can't do much because of my rib. Sunday, Gary baby-sits while Jaki and I help Dennis. Following week, we get wood to sawmill for milling. Dennis and I start roof with hemlock during this time and cut three or four trees in front of Cave. Get five hemlock and six spruce 2 × 6s from our own trees, which had to be cleared to let the winter sun get to the south wall. Depressing time, during which little gets done."

So it goes. On September 10, we helped the sawyer plane the pine, spruce, and hemlock 2 × 6s to uniform thickness and width. This is important in getting a good tight fit when roofing. In addition, the planed surface is a lot less prone to collecting cobwebs. A little wood (and strength) is lost, of course, but not enough to be significant. Our planking actually mea- sures 1¾" × 5¾".

8

Roof Deck

In planking, start at the bottom and work up. Get the first course of planking dead straight and make sure each end of the first course is equidistant from the peak. Snap a chalkline to mark the location of the bottom edge of the first plank. We had several long planks that were cut from the same logs as the rafters, so we were able to make the first course on each side of the roof from two planks each—a good start.

A word about lengths of planks and wastage: There are twelve rafters or eleven "spans" on each side of our house. A course might consist of two planks which will cover four spans and one which will cover three, for example. Each span is 32". However, the planks at the ends of each course must be long enough to accommodate an overhang. I figure an extra 12" for this, 10" for the overhang and 2" to get to the outside edge of the outermost rafters. (The 32" figure is measured center-to-center.) To minimize wastage, have the saw logs cut to useful lengths (Illus. 8–1). Take a copy of these lengths with you into the woods, or supply a list to the person cutting for you. Make sure that the logs are cut 2 or 3" *longer* than these

SPANS	LENGTH	SPANS	LENGTH
2	5'4"*	4	10'8"
LONG 2	6'4"	LONG 4	11'8"
3	8'	5	13'4"
LONG 3	9'	LONG 5	14'4"

* TOO SHORT FOR THE SAWMILL; CUT FOURS IN HALF

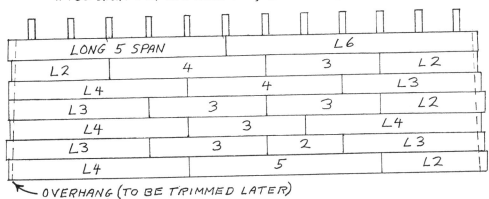

Illus. 8–1. To save wood, cut the logs to useful lengths.

Illustrations 8–2 and 8–3. If you use green wood, nail the planks as shown in these two photos, using a spike to keep them tight against each other, and nailing the lower nails first.

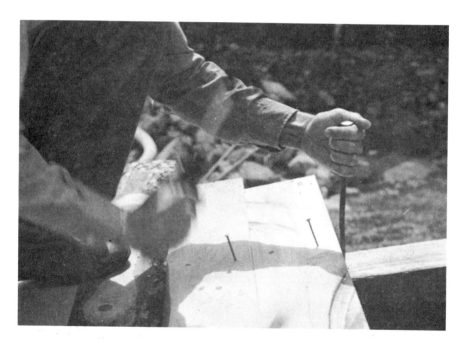

measurements, to allow for squaring the ends of the planks.

If you're stuck with green wood, as we were, nail the planks as shown in Illustrations 8–2 and 8–3. There's no reason why you can't draw 2 boards at a time. First, start the nails as shown. Then, take a star chisel or a 12″ spike and whack it about ⅜″ into the rafter tight against the plank.

Using the spike as a lever, draw the boards tight against each other and nail the lower nails first, then the upper ones. The reason for angling the nails is to make it easier for the wood to shrink without splitting (Illus. 8–4).

I like to use 16-penny resin-coated nails, though uncoated 16s would probably do. Resin-coated nails are hard to take out once they're in, but that's their purpose. The narrower shaft makes splitting at the ends of the plank less likely and makes the plank more likely to give with shrinkage.

Try to avoid placing too many joins one above the other (Illus. 8–5). Leave the planks sticking out over the ends. The overhang can be trimmed later by snapping a chalkline and cutting the whole edge at once with a circular saw or a chainsaw (Illus. 8–6). The chainsaw made cutting the edge easier. Green two-bys will choke all but the toughest circular saws. The only problem with the chainsaw is that you've got to be very careful to keep it perpendicular to ensure a straight edge.

Three things can happen at the peak (Illus. 8–7). One, the bottom sides of the opposing plank roofs may meet perfectly. In this unlikely case, sit back and relax. Two, they'll just miss meeting each other. This is what happened to us. In this case, make a

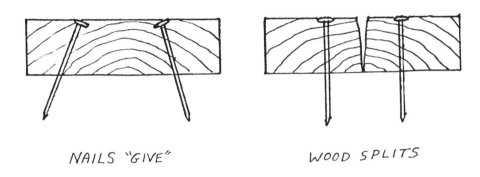

NAILS "GIVE"   WOOD SPLITS

Illus. 8–4. Angling the nails makes it easier for the wood to shrink without splitting.

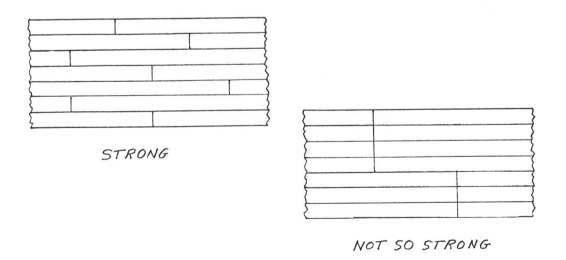

STRONG

NOT SO STRONG

Illus. 8–5. Avoid placing too many joins one above the other.

Illus. 8–6. Leave the plank ends sticking out. The overhang can be trimmed later by snapping a chalkline and cutting the whole edge at once.

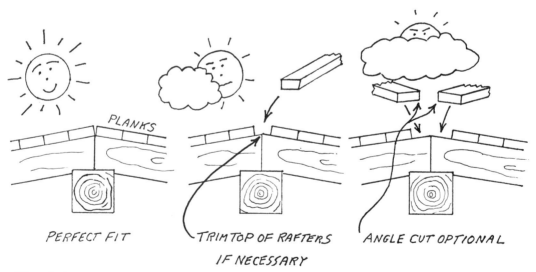

PLANKS

PERFECT FIT

TRIM TOP OF RAFTERS
IF NECESSARY

ANGLE CUT OPTIONAL

Illus. 8–7. At the peak, the planks might fit perfectly or just miss each other, or the last course on each side may be too wide.

wedge from leftover pieces and fit it into the space. You have to use a variable-angle table saw or good skill saw for this. Three, the last course on each side will be just too wide to fit. In this case, each of the last courses must be trimmed to the correct width. It's up to you if you want to incorporate the proper angle into the trim cut. The angle isn't necessary for strength, it's hidden from view by the girder, and it'll be covered by the waterproofing membrane.

The fascia or retaining boards serve to keep the sod roof from washing out over the gable ends of the house. The fascias frame the roof's high edges, directing all drainage down the slope. We used 2 × 12s, which worked out well. The 2 × 12s should be installed after the overhang is trimmed. Our method of fastening the fascia board involved first nailing a 2 × 6 plank right along the edge of the roof, for its full length, then nailing scraps of economy-grade 2 × 4s on top of the plank (Illus. 8–8). Together, the planks and scraps were now

Illus. 8–8. Fascia boards were fastened by first nailing a 2 × 6 plank along the edge of the roof, then nailing a scrap piece of economy-grade 2 × 4s on top of the plank. These two pieces provide a place for fastening the 2 × 12 fascia.

3¼" higher than the roofing planks, exactly the thickness of our rigid-foam insulation. At the original Cave, we placed particleboard over the insulation, then the waterproofing. It's much better for the insulation to be installed on top of the waterproofing, as we've done since 1979. The preferred method is described in chapter 9. The planks and scraps provided a high edge for nailing the fascia board. Use pressure-treated lumber for the fascias. After the fascia board was well nailed, 5/16" × 4" lag screws were used to make sure it was on permanently—a lag screw every 3' or so along the fascia.

It's a good idea to use a chainsaw to cut a groove 1" wide by ¼" deep down the middle of the bottom side of the fascia board before nailing it in place (Illus. 8–9). This groove serves as a "drip edge" and prevents rainwater from running along the underside of the roof planking and into the house.

Illus. 8–9. Use a chainsaw to cut a groove down the middle of the fascia board before nailing the board into place. This groove will serve as a "drip edge" to keep rainwater from running along the underside of the roof planking and into the house.

MOSS SODS

To stop earth from falling off earth roofs, I no longer use any form of retaining timbers, either fastened or laid loose. Taking a leaf from Wells's book, I now cut 6"-wide by 12"-long by 5"- to 6"-deep moss sods from the sand pit in front of Earthwood and lay them along the edge of the roof, kicking them tight against each other. After the earth or sod is installed, I water the moss thoroughly, and endeavor to keep it moist for a few weeks until the moss knits together. The edge of the roof is exposed earth for a while, but after a year or two, the moss completely spreads over the edge of the sods, creating an attractive border to

Illus. 8–10. These moss sods drain well, retain the earth, and look natural.

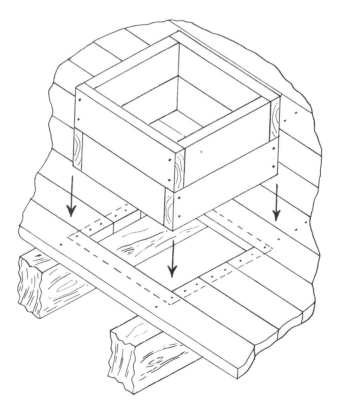

*Illus. 8–11. To accommodate skylights in the roof,
build the box frame on a flat surface and then nail it
around the opening. I recommend two courses of
2 × 8 material to build the box.*

the earth roof. Illus. 8–10 shows the moss sods on the edge of my workshop building at Earthwood. This detail is beautiful and natural, promotes good drainage, and stops the earth from falling off the roof.

SKYLIGHTS

Log End Cave had three skylights, one each in the office, living room, and bath. Now's the time to box in for the skylights. Follow the manufacturer's recommendations for sizing and have the actual skylight on hand to check against. The box frame that you need to make must be high enough to accommodate the 4″ (recommended) of extruded polystyrene insulation, the 2″ crushed-stone drainage layer, and the desired soil thickness. A 7″ soil layer seems to maintain the green cover in northern New York. These layers add up to 13″, so I recommend a 15″- or 16″-inch high box, composed, perhaps, of two courses of 2″ × 8″ wooden planks nailed together, as shown in Illus. 8–11. I found that the easiest way to install the box frame is to cut the opening in the roof with a chainsaw or circular saw, build the box frame on a flat surface, and then nail or screw the box frame around the opening.

Stovepipe holes can be cut in later, after the waterproofing membrane is applied.

9
Waterproofing

WATERPROOFING BEFORE INSULATION

In the correct order of events, now's the right time to apply the waterproofing membrane. The waterproofing should always be on the warm side (under) the rigid-foam insulation. This isn't what we did at the Cave, but it is what we've done on every earth roof since (six in number), and is the detail included in the 40' × 40' Cave plans. The potential problem with placing the membrane on the top (cold) side of the insulation is that warm moist internal air might travel through the planking and the insulation and condense on the cold membrane, causing a moisture problem. This is a common occurrence in older mobile homes. The cold metal roof is above the insulation. Internal moist air travels through the ceiling and, in the winter, reaches dew point on the underside of the roof, causing severe condensation. This is also why vapor barriers are always placed on the interior (warm) sides of insulation in standard stick-frame construction. At the Cave, we didn't actually encounter this condensation problem, but there's no doubt that we didn't get the order right. That insulation should be on the cold side of the vapor barrier is now accepted as fact by all leading authorities in the underground-housing field.

DON'T DO THIS!

Another change I've adopted since the late 1970s is my choice of waterproofing membrane. In 1977, based upon an old magazine article, I selected a method of inexpensive waterproofing. I wouldn't inflict this job on *anyone*: We built up our own homemade membrane, beginning with a trowelled-on layer of black-plastic roofing cement, available in 5-gallon pails. Next, we stretched on a layer of 6-mil black polyethylene (or *visquene*), pressing it into the incredibly sticky roofing cement. Then, to make doubly sure, and to establish beyond a doubt the masochistic side to my persona, I repeated the exercise: another layer of black cement followed by another layer of plastic. We used the same method on the walls, but limited the process to one layer each of roofing cement and plastic. All in all, we trowelled on 250 gallons of plastic roofing cement. One set of clothes and several pairs of gloves were destroyed during this process. All this happened during a three-week period when it seemed to rain almost every day, and Jaki's parents were visiting from Scotland.

BITUTHENE®

On every earth shelter since, I've used an excellent waterproofing membrane, Bituthene®. It's composed of two layers of 2-mil black polyethylene cross-laminated over a uniform (60-mil or $\frac{1}{16}$") thickness of rubberized asphalt. This bituminous bedding is very sticky, but the 36" × 60' roll comes protected by a layer of nonstick backing paper, which is removed while the membrane is applied. At Earthwood, we used Bituthene® 3000 membrane for both the roof and earth-bermed walls. There's also Bituthene® 3100 membrane for applications at temperatures between 25°F and 40°F.

WATERPROOFING THE WALLS

On a building like Log End Cave, where the berm comes right over the east and west walls and melds with the earth roof, waterproof the walls first, simply because the roof membrane should overlap the wall, and not the other way around.

After the surface-bonding cement is cured, the wall is primed with Bituthene® 3000 primer. The primer goes on very quickly with a roller and dries in an hour.

Before applying the membrane, we attend to the critical waterproofing detail where the wall meets the footing. A strip of Bituthene® about 7" wide and 36" long is cut from the end of a roll. The backing paper is removed and the strip is carefully fitted into the crease where the wall and the footing meet. About 3" of the strip is now stuck fast to the footing, and the remaining 4" of the strip is stuck to the rest of the wall. The edge on the footing is sealed for protection against "fish-mouthing" (raising of the membrane) by the application of a bead of Bituthene® EM3000 mastic. This bead is applied with a caulking gun (Illus. 9–1) and then

smoothed with a knife. All freshly cut edges of the membrane are sealed in this way, but it isn't necessary to apply mastic to the factory edge, which comes already protected by a factory-applied mastic along it.

Sheets are applied vertically, as shown in Illus. 9–2. Cut the sheet to the right length, and have someone hold the top edge from above. On the east and west walls of the Cave-type design, this membrane would begin 6" onto the roof, or about at the center of the first course of planking. Two other people work from below, one removing the backing paper, the other pressing the Bituthene® into the wall. Lap the previously laid strips at the footing detail with a full 3" overlap, and seal the bottom with mastic. To prevent bubbles forming behind the membrane, adhesion is best begun at the middle of the sheet and spread out to the right and left with your hands. Continue pressing the Bituthene® into place right up onto the primed planking, and discard the backing paper. The second 36"-wide piece laps the first by 2½". A yellow lap line is provided on each edge of the sheet to aid in the placement of the next sheet. If it becomes

Illus. 9–1. The important detail where the footing meets the wall is sealed with a strip of Bituthene®.

Illus. 9–2. The Bituthene® waterproofing is applied vertically to the walls.

apparent that the second piece is wandering off line, stop pressing the Bituthene®. The sheets can't be pulled or stretched back to alignment. Cut the sheet short at this point, and begin again with an overlap of at least 3″ at the bottom. Use heavy hand pressure at the lap seam and seal the newly cut edge with the compatible mastic. Never use an incompatible mastic such as tar, asphalt, or pitch-based materials, or mastics containing polysulfide polymer.

The advantages of Bituthene® are: speed, ease of application, and quality control. Any shape piece can be cut to fit, but always allow for the 2½″ or 3″ lap and seal all cut edges.

Happily, the Bituthene® membrane sticks extremely well to aluminum flashing. If the 2″ × 12″ retaining timber is used, the detail would look like that shown in Illus. 9–3.

With the moss-sod method, just flash the edge of the planking with either an aluminum drip edge made for the purpose or a piece of 10″-wide aluminum flashing bent at a right angle along a straight line, so that 7″ is nailed to the deck and 3″ hangs over the planking to form a protective drip edge. The Bituthene® sticks to the nailed part of the flashing, covering the nails. Keep the Bituthene® membrane about 3″ in from the edge so it's not exposed to the sun's ultraviolet rays. As long as it's

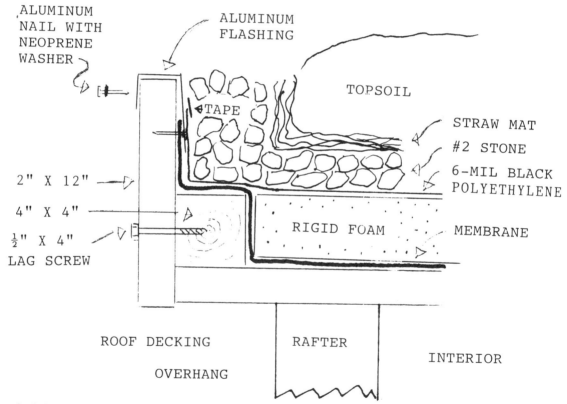

Illus. 9–3. New retaining-board detailing for parapet walls, such as the gable ends of a Log End Cave design. You could use moss sods instead of retaining timbers.

WATERPROOFING THE ROOF

Flashing is one of the most important jobs in waterproofing and one of the most difficult to do well.

buried under the earth, the black-plastic surface will last forever or 100 years (whichever comes first).

The Bituthene® goes down quickly on a simple

roof plane like that of Log End Cave. The application is horizontal, beginning at the bottom of the roof, and lapping a good 3″ onto the membrane already applied while waterproofing the walls. Roll out the approximately 30′ of membrane across the roof with the backing paper still on. Cut the membrane to the right length and establish the 3″ lap onto the wall membrane which comes up onto the roof. Now, pull up about 3″ of the backing paper along the width of the membrane at the starting end and stick the membrane down onto the roof-edge flashing, making sure you've still got that 3″ lap at the low side of the sheet. Roll up the sheet again. Now *unroll* it again, this time while someone is pulling the backing paper out from under the sticky bitumastic. The membrane should fall right into place. But if it doesn't, if it looks as if you're wandering off the yellow guideline, stop. Cut the sheet short, lap back 3″, and try again. If care is taken, and the technique given in this paragraph is adhered to, you'll never have to cut short and start again with the 3″ lap, but you should know what to do, just in case.

Punctures, cuts, or other suspicious marks on the membrane are easily remedied with a patch cut from the same material. The patch should extend 3″ in every direction from the damage, the patch pressed hard to the surface with the heel of your hand. The edges of the patch are sealed with mastic. To diminish the chance of a corner lifting, I make octagon-shaped patches with a razor-blade knife.

FLASHING SKYLIGHTS & CHIMNEYS

Illus. 9–4 shows the aluminum flashing of a skylight at Log End Cave. Note that the flashing is tucked *under* the black plastic on the uphill side (upper right of the picture) and *over* the membrane on the downhill side. With the Bituthene® membrane, I install the flashing a little differently. I install it over the membrane *all around*, compressing a bead of Bituthene® mastic with the nails. Then I cut a 6″-wide strip of Bituthene® and lap it 3″ onto the flashing (covering the nails), and 3″ onto the surrounding membrane. Again, caulk cut edges with mastic. The corners close to the skylight box are the critical details. Here, use plenty of

Illus. 9–4. Flashing at the original Log End Cave was under the plastic membrane on the uphill side and over the membrane on the other sides. This flashing detail is much easier now when using the Bituthene® membrane.

Illus. 9–5. The Metalbestos® flashing cone is set in place over the Bituthene®.

Illus. 9–6. After nailing, the edges of the flashing are sealed with strips of Bituthene® and caulked with a compatible mastic.

overlapping Bituthene®, small pieces covered by larger pieces. The last (topmost) layer must have the edges caulked. When the earth goes on, surround all projections like chimneys and skylights with a 6"-wide crushed-stone drainage buffer, the crushed stone meeting the underground drainage layer described in chapter 11.

I'm a great fan of the Metalbestos® chimney system for use with wood stoves, and such a system works particularly well with earth roofs. The stovepipe itself is stainless steel inside and out, and sandwiches a 1" noncombustible layer of low conductivity. The pipe is rated safe within 2" of combustible material. There's no asbestos in Metalbestos®. The beauty of the system with earth roofs is that the pipe manufacturer makes a tall aluminum flashing cone suitable for shallow-pitch roof applications. The cone comes with a nice wide flange, which makes it easy to use with Bituthene®.

Cut out a hole in the finished roof with a diameter 4" greater than the outside diameter of the pipe you'll be using. A chainsaw works well for cutting this hole. Hang the whole chimney through the hole with a piece called a "Roof Support Package" (RSP), which screws to the deck, and add at least one piece of chimney above the roof. Lower the flashing cone over the chimney and use 6" strips of Bituthene® to seal the edges of the flashing-cone flange, the same as with the skylight (Illustrations 9–5 and 9–6).

At Earthwood, a 3'-diameter masonry chimney rises up out of the center of the house. I had a special flashing cone made at a sheet-metal shop, and nailed the cone through the Bituthene® to the deck, using ribbed nails every 2", causing extra

Illus. 9–7. A galvanized flashing cone was specially made for this cylindrical chimney.

mastic applied around the edge to ooze out. The nails compress a neoprene washer to seal off water penetration. The top of the flashing cone is sealed with silicone caulk (Illus. 9–7).

While there are other waterproofing options, I don't hesitate recommending the Bituthene® membrane to owner-builders because of its ease of application, quality, and moderate cost.

10

Insulation, Drainage & Backfilling

EXTERIOR-WALL INSULATION

It's convenient to apply the wall insulation before installing the footing drains. In northern climates, I now use 3" of extruded polystyrene down to frost level (48" in our area) and 2" the rest of the way down to the footings. This wall insulation should cover the top of the footing as well as the insulation which was placed on the outside of the footing tracks. Try to prevent any direct conduction from the interior to the exterior sides of the concrete fabric of the building. Malcolm Wells calls any such conduction an energy "nosebleed," and, like a drippy nose, the manifestation can be condensation on the inside, the result of dew point occurring on the inner surface.

At Log End Cave, we used 2" of beadboard for the first 48" of depth, and 1" from that point down to the footing. We didn't wrap the footings with insulation, and suffered the worst manifestation of energy nosebleed: condensation at the base of the wall in late spring and early summer. Beadboard (*expanded polystyrene*) may be okay to use with good drainage. Use a dense form of beadboard, given a choice, as opposed to the very light and easily compressible light white stuff which is commonly sold. Research the R-value and the price. The denser beadboards will have higher R-values, some of them equivalent to extruded polystyrene's R-5 per inch. You should be concerned with R-value per dollar. If there's little difference between the cost of expanded and extruded, go with the extruded, which is close-celled and will stay dry and crisp underground until the end of time.

We used a fairly low-density beadboard, but even today, 17 years later, it's still in good condition. I credit the good drainage provided by the sand backfill as being the reason for its preservation. I've seen beadboard completely saturated, and in that condition it has no value as insulation.

I don't recommend polyurethane foams, which are generally yellow in color and often have one or two layers of aluminum foil on the surfaces. Although the starting insulative value is very high at R-8 per inch, the foam is prone to water absorption, which can bring that value right down to R-3 or less.

On nice straight walls, like those in the Cave, it's easy to spot-glue the rigid foam to the walls. Six dollops of compatible mastic should hold a 4' × 8' sheet in place until the wall is backfilled. There are actually glues made for rigid-foam insulation, available for convenient use with a caulking gun. Don't glue the insulation too early, as we did. After a few days of wind and rain, our beadboard was falling off the wall. Glue the insulation just before backfilling. At Earthwood, with its round walls, we had to manually hold the Styrofoam® against the wall during backfilling, because no glue will hold the foam against the wall if the wall is even slightly curved. Remember, it's called *rigid* foam.

DRAINAGE MATERIALS

Waterproofing was discussed before drainage because the former's installation occurs first. Drainage is your main line of defense against water penetration. Give water an easier place to go than into your house, and it will cooperate. The keys to good drainage are the use of backfill with excellent percolation characteristics, and a footing drain (*French drain*), which carries the water away from the house. At the Log End Cottage basement, we installed footing drains correctly, but backfilled with the poor claylike earth which came out of the hole. The water couldn't get to the footing drains. Hydrostatic and frost pressures caused a movement in the block wall, negating our cement-based waterproofing. Cement waterproofings aren't recommended, because they lack the ability to bridge even the tiniest shrinkage or settling cracks.

A good drain system for an earth-sheltered house like the Log End Cave is shown in Illus. 10–1. With a freestanding roof, such as at Earthwood, a plastic or aluminum gutter is advisable, particularly in wet climates. Using gutters eliminates the need for the topmost drain, the one just below the surface. The intermediate drain is optional, and is only necessary in soils with moderately good percolation characteristics. If good percolating backfill (such as coarse sand or gravel) isn't available at the site, it should be brought in. Backfill should be compacted in 12″ layers. Using 12″ layers minimizes the tamping pressure necessary, decreasing the likelihood of damage to the walls.

If no good backfill is available within hauling distance, several horizontal drains may be necessary. Another option is Enkadrain®, a fibrous ½″-thick mat (manufactured by AKZO Industrial Systems Co.) which is placed against the wall. Hydrostatic pressure is eliminated, because water entering the tough nylon mesh is rapidly carried down to the footing drain. Enkadrain® isn't a substitute for waterproofing; it's a complement. The 0.4″ thickness is suitable for depths of up to 10′ below grade and was about $0.50 per sq. ft. in 1991. In some cases, using Enkadrain® may be cheaper than hauling in good backfill.

Another good product, Dow's Styrofoam® Therma-Dry®, combines insulation and drainage in one application. An abundance of vertical grooves (¼″ deep × ⅜″ wide) are cut into the outside of a 1½″- or 2¼″-thick sheet of Styrofoam® Blueboard®, and covered with a nylon filtration mat. Water seeps into the grooves and is carried

down to the footing drain. As insulation and drainage are combined in one product, the cost of $1.60 per sq. ft. (1991) is quite favorable.

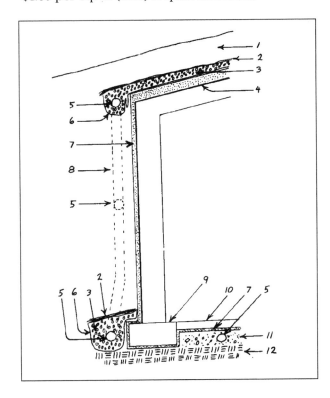

1. Earth.
2. Filtration mat; can be hay or straw.
3. Crushed stone.
4. Waterproofing membrane.
5. 4″ perforated drain wrapped in a filtration fabric.
6. 6-mil polyethylene to form underground "gutter." The top gutter should include a fold in the plastic as shown, to allow for earth settling.
7. Rigid-foam insulation.
8. Vertically placed 4″ nonperforated drains connect the horizontal perforated drains. T-junctions are available to make these connections.
9. Footing.
10. Concrete-slab floor.
11. Compacted sand, gravel, or crushed stone.
12. Undisturbed subsoil or heavily compacted pad, as previously described.

Illus. 10–1. A good drain system for an earth-sheltered house. (Not drawn to scale)

Illus. 10–2. The footing drain should be perforated-plastic flexible tubing, laid adjacent to the footing on a bed of washed 1″ stone.

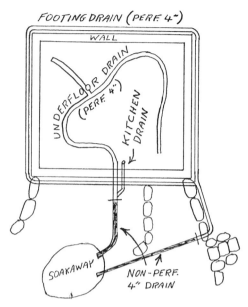

Illus. 10–3. The underfloor drain, kitchen drain, and footing drain all take separate paths to the same soakaway.

Illus. 10–4. From the southeast corner (the lowest point of the footing drain's grade) the pipe will be run at a similar slope behind the retaining wall and into a soakaway.

Illus. 10–5. Hay spread on top of the stones covering the footing drains will decompose and form a mat to stop infiltration into the drain by the subsequent sand backfill.

FOOTING DRAIN

The other key to good drainage, besides good percolation, is a good footing drain.

The drain is constructed of 4" perforated flexible plastic tubing, preferably covered with a nylon or fiberglass "sock" to prevent infiltration of silts, which could clog the system. The tubing is laid out on a 3" to 4"-deep bed of washed 1" (#2) crushed stone adjacent to the edge of the footing (Illus. 10–2). The drain should slope slightly as it travels around the perimeter of the house and into the soakaway, or (better, if it's possible at your site) out above grade (Illus. 10–3). Our method of establishing this slope was to make marks with a crayon 4" below the top of the footing at the northwest corner, 5" below the northeast and southwest corners, and 6" below the southeast corner. We joined the marks by snapping a chalkline and brought in washed #2 stone, bucket by bucket, to the grade established. The slope, then, ran in both directions around the house from the high point in the northwest corner to the low point in the southeast corner, about an inch of drop for every 30'. From the southeast corner, shown in Illus. 10–4, the perforated pipe ran at a similar slope behind the location of the large retaining wall and on to a soakaway. (There's no reason why the retaining walls couldn't be built at this time, but we were in a hurry to get the sod roof on before winter, so the retaining walls were delayed.) The footing drain is well covered with additional crushed stone, about 3" above the top, and the stone is covered with 2" of hay or straw (Illus. 10–5). The hay partially decomposes and forms a filtration mat which keeps the crushed stone clean. Although manufactured filtration mats made for this purpose are available, this "organic" filtration technique works well.

BRACE THE WALLS

The footing drains completed, the only preparation before backfilling was to brace the walls from the inside as a safety measure against the weight of the sand that would be dumped against the sidewalls.

Think of a below-grade block wall as a beam resisting a load. In this case, the load is lateral. Earth pressure comes from the *side*. Because bending stresses increase as the square of the span of a beam (the distance between corners in the case of the wall), shortening the span greatly increases the resistance of the wall to lateral pressure. Cutting the span in half increases the bending strength by four times. Internal braces are one way of reducing the span. Built-in block pilasters are another means of resisting the pressure, but they're cumbersome and difficult to work in with the floor plan. I prefer to use 12" blocks and eliminate the pilasters. And, of course, there are bond-beam courses, as already described.

At Log End Cave, bracing against lateral pressure was easy, since we'd already framed most of the internal walls with 2 × 4 framing, and all we had to do was nail a diagonal brace to the framing for support. We braced at every point around the perimeter of the house where an internal wall met a block wall (Illus. 10–6).

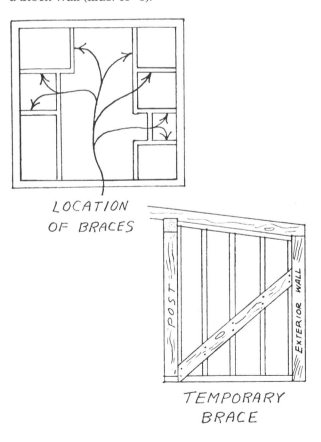

LOCATION
OF BRACES

TEMPORARY
BRACE

Illus. 10–6. Brace the frame from the inside to support it against the weight of the sand that will be poured against the sidewalls.

Never backfill a basement foundation or underground house before the tops of the block walls are tied together, either by the floor structure, or, in our case, by the roof-rafter system. These internal frameworks provide valuable additional resistance against lateral loading.

BACKFILLING

At Log End Cave, we brought in 25 dump-truck loads of sand (about 125 cubic yards) to backfill the walls, because we'd learned to our cost at the Cottage basement that the subsoil at Log End doesn't have good percolation qualities.

Backfilling can be done with a bulldozer, backhoe, or front-end loader. Using the front-end loader attachment for a farm tractor would be a good choice. Maybe a neighbor can save you some heavy-equipment costs.

During backfilling, be alert to the rigid foam moving or coming away from the wall, and watch for big stones which might come crashing down against the wall. Since the equipment operator might not be able to see such happenings from his perch high above, another pair of eyes (yours) is imperative. Note that the placement of the rigid foam over the membrane protects the membrane from damage, within reason. The foam won't necessarily save the wall from a 500-pound boulder.

11

Earth Roof

THEORY

When I teach underground housing at Earthwood Building School, there's always the studious type who listens very carefully, takes good notes, and arrives at some very logical questions.

"If earth is a poor insulator, then why bother putting any on at all? I mean, you'd need three feet of the stuff to be worth anything. Why engineer and build for all that load? And why risk the leaks?"

I usually reply, "Good question, and it deserves a good answer."

As I begin my reply, I write a list of seven single words. Yes, earth is poor insulation, and, yes, 3' of earth is nutsy. Yes, such a heavy roof can be built, and some have been built, but what did houses with such roofs cost? What kind of technology was used? Pre-stressed concrete planks installed with a crane? Not post-and-beam, that's for sure. We shouldn't cut the kinds of trees that would be needed to support a load of 500 pounds per square foot. The right amount of earth to have on the roof is just the amount needed to maintain a green cover, no more, no less. In a temperate climate with moderate rainfall, 6" to 8" of earth seems to work just fine. If the roof dries out, as the Earthwood roof did once during a long summer drought, it'll come back, maybe with some surprises. It's amazing the new plants you'll discover on your roof. Here are the advantages of having an earth cover:

- **Insulation.** Earth isn't completely without insulative value, but by far the greatest such value is found in the first 3" or 4". The soil is rich and loamy, aerated by the root systems of the grasses, and not greatly compacted. The grass itself, especially if allowed to grow long, supplies additional insulation. An earth roof holds snow better than any other kind of roof. A house that has snow on the roof where there's heat inside is a warm house. Light fluffy snow adds insulation at R-1 per inch of thickness. A two-foot snowstorm of light dry snow adds R-24 to your roof. We certainly found, while living at Log End Cave (and at Earthwood), that houses are easier to heat when they have a good layer of snow on top.

  It's true that a house built in our 9000-degree-day climate with nothing more than 7" of earth on the roof wouldn't work. You still need the rigid foam, at least R-20's worth. If building-code officers insist on R-38, those first few inches of grass and roots and loose earth and snow will make up for the shortfall. Remember, too, that the R-38 derived from light fluffy insulation contributes nothing to the thermal capacitance of the building. An earth roof is grass *and* mass.

- **Drainage.** The conventional roof doesn't slow runoff during heavy rains. A lot of water has to be dealt with quickly. In a bermed house, water percolates slowly to the footing drains. Much of the water just runs naturally away from the building, as water runs down any hill. Hundreds of gallons of water are absorbed by the earth, which keeps it in reserve to replenish the atmosphere on a dry day, which brings us to . . .

- **Cooling.** Water evaporating off the earth roof in the summer promotes cooling. Even a few inches of earth protects the substrate from the high surface temperatures found on an asphalt or metal roof. You can cook an egg on an asphalt roof. Stick your finger in a wet lawn, however and it comes out cool.

91

- **Aesthetics.** One of the main reasons for building an earth shelter is its minimal visual impact. An earth-sheltered home harmonizes with its surroundings. If planning and zoning officials really cared about the visual amenity within their jurisdiction, they'd insist upon underground houses . . . with earth roofs.
- **Ecology.** The earth roof supports life and oxygenation. Save 1500 sq. ft. of the planet's surface from being converted to lifeless desert. Wildflowers and leafy vegetables can be grown on a shallow earth roof. Forget the shrubs and trees.
- **Protection.** Yes, a massive earth roof is better protection against sound and radiation than a 7″ earth roof, but again, as with insulation, the first few inches have the major impact.
- **Longevity.** Do it right the first time. Read this chapter carefully and follow its advice, and you'll never have to do it again. There are two things that break down most roofing materials: ultraviolet radiation and freeze-thaw cycles. With an earth roof, the sun's rays never get to the waterproofing membrane, so there's no UV deterioration. I've dug up black plastic buried for 25 years and it was just as pliable as the day it was buried. Since the 1950s, Americans have been preserving their garbage for future posterity wrapped in nonbiodegradable plastic bags.

Before I can continue, I'm interrupted by another question, this time from an environmentalist.

"I'm very concerned about using all these petrochemical products and burying them where they'll never break down. Could you speak to that?"

Gladly. The properly built underground house will *save* fossil fuels in the long run. More efficient use of energy is still America's best potential for reducing dependence on fossil fuels. There are European countries with standards of living equal to ours, but with only half the energy consumption per capita. And some of these, like Sweden, have long, cold winters. At Earthwood, we use less than half the energy for heating and cooling as other homes of comparable size in our area use. Polystyrene insulation, which saves fuel, is a better use

of petrochemicals than burning the extra fuel oil required without it. And now, with the advent of replacements for the chlorofluorocarbons (CFCs) heavily used as expanding agents in the past, the ozone layer needn't suffer further damage from rigid-foam insulation.

Nonbiodegradable membranes and insulations? For a long-lived house, this is exactly what we want. We don't want something which is going to break down and pollute the earth and water table. We don't want something that has to be replaced someday, because it's difficult to replace insulation or waterproofing in an underground house.

In summary, the main gain from putting 3′ of earth on the roof instead of 1′ is that the entire structure is set 2′ deeper into the ground. At soil depths of 6′ to 12′, there's an average difference of 1°F for each foot of depth, so setting the home 2′ deeper amounts to a 2°F advantage, both for summer cooling and winter heating, plus the slightly greater insulation derived from the extra 2′ of earth. It's my contention that an extra inch of rigid foam (which weighs practically nothing) is a much more reasonable means of achieving the same advantage than is building the behemoth required to support a 3′ earth load. All we should be concerned about is maintaining the green cover. For us, 6″ to 7″ of earth has worked out well. The roofs have dried out in droughts, but they always bounce back. Don't water them. Don't mow them. In areas of low rainfall, 10″ to 12″ of earth might be better to promote a deeper root system. Find a grass with good ability to bounce back from extended dry periods.

INSTALLING RIGID FOAM

On the roof, use either extruded or high-density expanded polystyrene. You're going to install a drainage layer over the rigid foam anyway, so your main concern should be R-value per dollar. Get prices locally and figure out the best value.

In our climate (9000 degree days), I like a minimum of R-20 on the roof (equivalent to 4″ of extruded), but R-25 (5″) would be nice if you can afford it. The earth, grass, roots, snow, etc., will

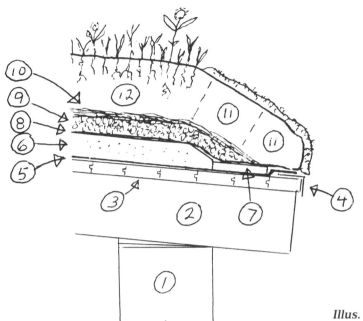

1. Above-grade wall
2. Heavy wooden rafter
3. 2" X 6" T&G planking
4. Aluminum flashing as drip-edge
5. Bituthene® (or equivalent) membrane
6. 4" to 5" rigid-foam insulation
7. 1" rigid foam or ½" fibreboard to protect membrane
8. 6-mil black polyethylene
9. 2" of #2 crushed stone drainage layer
10. Hay or straw filtration mat
11. Moss sods cut from sandy soil
12. 7" to 8" topsoil, planted

Illus. 11–1. Roofing detail for a freestanding earth roof using moss sods to retain the earth

bring the roof up to code values, and you still have the advantage of building in an ambient (the earth) which is 20 to 60 degrees warmer in winter than the outside air enveloping a surface-built house.

Installing foam is easy. Just lay out the foam in large sheets. Four-by-eight-foot sheets are easier to work with than twice as many 2' × 8' sheets. Use 2"-thick pieces, as opposed to the very light 1" sheets, which are more prone to being carried away by the wind during installation. Tongue-and-groove foam sheets are nice, if they don't cost more money. Otherwise, lapping the first course with the second will do the same job as T&G. In case the sheets become separated, lap the first course with the second, even if T&G is used, to guard against energy nosebleed. Lots of duct tape helps. Make sure you get good overlap onto the sidewall insulation.

It isn't necessary to insulate the overhang of the building, but I like to put an inch of foam there as a protection layer over the membrane. This foam stops stones from putting a concentrated load on the membrane. The edge detail of a freestanding

earth roof using moss sods to retain the soil is shown in Illus. 11–1.

On Earthwood's 4" roof insulation, which was made in two 2" layers, we tacked the second course to the first by pushing in hundreds of 16-penny (3½") nails. These nails were for wind protection until all the insulation was applied. As you lay out the plastic described in the next paragraph, you can pull out the nails. I don't like the idea of hundreds of nails aiming towards my waterproofing membrane, although the nails are still ½" away from doing any damage. These good and valuable nails can be reused.

DRAINAGE LAYER

We create a drainage layer under the soil in three steps.

• Spread a sheet of 6-mil black polyethylene over the entire roof. Look again at Illus. 11–1. This layer is going to shed most of the water to the edge of the building, taking the pressure off the waterproofing membrane. I consider this inexpensive insurance layer to be the best-spent money in the whole project.

- Using 5-gallon buckets, spread 2″ of #2 crushed stone on the plastic. Now you can relax, because a hurricane won't blow your work away.
- Over the crushed stone, spread 2″ of loose hay or straw, creating a filtration mat to keep the drainage layer clean, or buy a product made for the purpose. Environmentally, I like hay or straw.

SOD ROOF

At Log End Cave we actually cut sods from the corner of a field which we had tilled, de-stoned, and planted with timothy and rye back in June. We found that 2″ or 3″ of topsoil came up with each cut sod. As this was insufficient depth of cover to maintain the grass, we spread 4″ of good topsoil all over the roof prior to installing the sods. Remember that we were putting earth directly onto the membrane at Log End Cave, a practice I no longer employ, so we screened out stones from the first inch of soil. Such screening is unnecessary when putting soil straight on the hay or straw filtration mat. Remove any stone larger than a tennis ball, as such stones conduct heat and don't contribute to the soil from the point of view of retaining moisture or supplying nutrients for the green cover.

We put on the 4″ of topsoil in late October. A friend experienced in golf-course maintenance advised me that it was too late in the year to put the sod on. I was inclined to agree, but wanted to do something to approximate the effect of a 6″ earth roof in order to test the heating characteristics of the home during the winter. I also wanted to know for sure what *would* have happened if we'd put the sod on at the end of October. On most of the roof, then, we spread a 2″ layer of hay to approximate the insulative value of the missing sod and to prevent erosion of the 4″ of topsoil during the spring thaw. We spread pine boughs over the hay as protection against the wind, and weighted the whole mess down with long, heavy sticks.

As a test, I cut and laid 20 sq. ft. of sod near the peak of the roof. My friend was right. The test patch was still in bad shape late the following June. The rest of the roof, sodded in early June, was flourishing and in need of a second mowing. (At Log End, the earth roof was an extension of the lawn area around the home, so we kept it mowed. At Earthwood, the roof is "freestanding," and we rarely mow.)

Sodding was hard work, but three of us, one cutting, two hauling and laying, applied the sod in two days (Illus. 11-2). With the east and west berms, this amounted to about 1600 sq. ft. We found that 10″-square sods were the best size; larger pieces would break up in transport. Sod should be cut damp, but not soaking, so that it will hold together and establish itself in its new location.

We accidentally discovered what may be the easiest way of all to start an earth roof. The hay we'd used as insulating mulch was full of seeds, and the roof was becoming quite green on its own in the spring. Unfortunately, we still needed another 2″ or 3″ of earth to maintain a green cover. Had we known, we might have applied the whole 6″ of soil in the fall, and let the hay take over naturally in the spring.

PLANTED-EARTH ROOF

We found out at Earthwood that it's easier just to spread all the topsoil required (Illus. 11–3), and then plant it. The "topsoil" was a silty material which came from a stream's floodplain on a nearby farm. The soil is free of stones and has relatively poor percolation, good for this purpose, as less earth is needed to retain water for use by the plant cover. We used 8″ of earth, which eventually compacted to 7″. The drainage layer takes care of oversaturation. We planted rye in early September and, thanks to a beautiful autumn, grass was coming through the mulch layer within weeks. By late October the grass was thick, green, and lush.

An alternative to grass is a roof of wildflowers. Vetch, purple and white, works well, as do many other flowers. Seek local advice and choose flowers suitable for the soil, climate, and sun/shade ratio of the roof.

One final point: If you decide to use heavy timbers such as railroad ties or pressure-treated landscaping timbers to retain the earth roof, be sure to include three details shown in Illus. 11–4.

Illus. 11–2. Laying sod is easy: With the side of your foot, simply kick each new piece against those already laid.

Illus. 11–3. Jaki spreads the topsoil using a 5-gallon bucket.

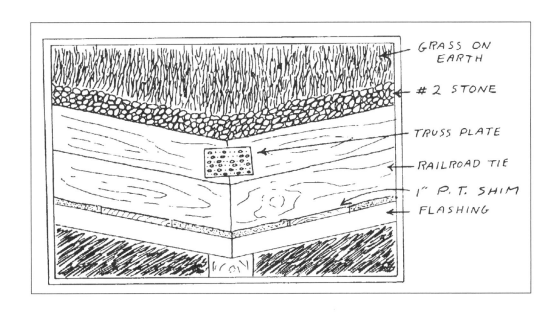

Illus. 11–4. Retaining timbers are truss-plated to each other.

(1) Use galvanized truss plates to fasten the timbers to each other. (2) Provide good drainage next to the timbers by the use of crushed stone right down to the main drainage layer. This drainage will stop lateral pressure on the timbers. (3) Use 1" or ⅝" pressure-treated shims to keep the timbers an inch off the roof substrate. Shimming prevents ice-dam buildup, a problem we encountered at Earthwood and cured by retrofitting said shims later. It's easier to do this shimming at installation. Try the moss sods. That's what I'll be doing from now on.

12

Closing In

Closing in Log End Cave involved four different tasks: installing windows, building and hanging the door, installing vents, and infilling with cordwood masonry.

There were seven windows at the original Cave, not counting three skylights or the small Thermopane® in the door. Each of the seven is a 1"-thick insulated-glass unit composed of two pieces each of ¼" plate glass enclosing ½" of an inert gas (such as nitrogen). All the windows were custom-made for the job by a local firm specializing in insulated windows.

At the design stage of your house, go to the nearest manufacturer of insulated glass and ask them if they have any leftover units. All these firms seem to store perfectly good clear units which for some reason were never picked up by the customer. Maybe they were cut the wrong size. Maybe the customer went bankrupt. The upshot is that many manufacturers are happy to sell you these units at a fraction of their normal price. It's actually cheaper for them to sell the units at 50% to 75% off than to take them apart, clean the glass, and remake a smaller unit from them. Hard to believe, but true. There are a surprising number of these small companies in business; find them in a classified telephone directory.

The three large south-facing windows at the Cave were found this way, and they were such a bargain that I designed the house around them. Imagine my chagrin, then, when I went to pick them up only to find that the owner of the plant had sold them to someone else after he'd agreed to sell them to me. To his credit, he made it up to me by making me brand-new units to fit, at a very good price. All seven windows (plus the little window in the door) cost a total of $322.50, in 1977 dollars.

The safest course is to buy them and store them, as shown in Illus. 12–1.

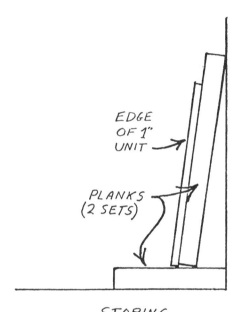

STORING
WINDOW UNITS

Illus. 12–1. Keep the window units on planks in a safe place while marking the sills and molding for installation.

Make your openings ½" larger in each direction, so that you can leave a ¼" dead-air space around the entire unit. A 48"-square unit needs an opening of 48½" × 48½". If you're going to order the units to fit, it's probably best to order them after the rough opening is actually framed. This way the *actual*, not the *theoretical*, dimensions can be quoted. Don't forget to subtract the ½" from each measurement.

I wouldn't advise single-pane windows at all, at least not in northern climates. The condensation problem would be horrific in the winter, and the heat loss entirely out of keeping with a home which is supposed to be energy-efficient. Even acrylic skylights should be double-pane, as ours were. You can even buy double-pane skylights that crank open in the summer, with a screen inside to stop insects. In the winter, the screen can be replaced with an internal storm pane, to further decrease heat loss.

A double-glazed window properly installed on the south wall of a house actually admits more heat than it loses, even if shutters or insulated drapes aren't used. Unfortunately, even insulated-glass windows are big heat losers when installed on the north side of the house. Here, triple-panes might be cost effective. At Log End, we shuttered off the north-facing windows during the winter, with little loss of light, as the snow usually covered them anyway, and the sun was low enough in the sky to penetrate deep into the back of the home. The skylights gave a lot of light, if you went up on the roof once in a while to brush the snow away.

We installed the three large south-facing windows first. The procedure follows the three steps described below.

INSTALLING EXTERNAL MOLDING

We used rough-cut 1″ × 1″ stock for window molding, inside and out. Aesthetically, such stock is in keeping with the rough-hewn style of the house, and it's cheap. Our stock was a product of our hemlock roof, although I prefer pine, which splinters less. We installed the topmost piece first, towards the exterior of the 4″ × 10″ lintel so that we'd have useful wide shelves on the inside.

I cut the molding to length and set 8-penny cup-headed nails so that they were just starting to break through the surface on the other side. Nowadays, I use my reversible drill and screw the molding in place. Using screws not only permits easy withdrawal, but there's less chance of smashing a window. I run a bead of silicone-based caulking along the edge where the nails (or screws) are starting to break through and fasten the piece in place. I like to lay the molding up against a known straight-edge guide to be sure that the molding is

installed straight. The caulking compresses, giving a permanent seal against draft. A word about caulking: Silicone is good, maybe the best, but it's expensive. A more economical (and still high-quality) alternative is to use acrylic-and-silicone hybrid caulks.

Using gloves and the aid of two helpers, I put the window in place against the top molding, make the window plumb, and mark the sill with a pencil run along the outside edge of the glass; then we put the unit back in a safe place for a few minutes.

The bottom molding can now be installed, followed by the two vertical side-molding pieces. I don't drive the nails all the way home, just in case an adjustment is needed. You can check again with the window unit itself. I can't think of any other way to check that the four pieces of molding are all in the same plane. If the window fits flush against all the molding pieces, we remove the unit and drive all the nails home. A bead of caulk is run around the inner perimeter of the molding so that the glass will compress the bead when the unit is permanently placed.

SETTING THE WINDOW IN PLACE

Set the window on ¼″ wooden shims, leaving a dead-air space between the unit and the surrounding frame. Two or three shims will suffice for a 7′-long window. Rubber shims are also made for the purpose; perhaps you can get some where you got the glass. The shims are 1″ or 2″ long and just a whisker less than the thickness of the unit, so that the plate glass itself will rest on the shim without the shim getting in the way of installing the internal molding. Now carefully place the unit on the shims and press firmly to compress the bead of caulk. The unit is held in place with a small scrap of molding tacked at the middle of each end of the window. Now do the internal molding.

INSTALLING THE INTERNAL MOLDING

This is pretty much the same as the procedure just described, except that the window unit doesn't have to come up and down anymore. Start with

Illus. 12–2. Double-caulk the windows by running a bead of silicone along the glass itself and another bead along the molding surface to be placed against the sill or lintel.

Illus. 12–3. Installing the trapezoidal window in the north wall. Since it was convenient to work from the outside, I fastened the inside molding first.

the top and bottom strips of molding. Again, caulk against the glass and between the molding and the frame. One way to do this double-caulking is to run a bead along the glass itself and, after the nails have been started, run another bead along the molding surface which is to be placed against the sill or lintel (Illus. 12–2). Nail the molding, taking care not to smash the window as you drive the nails home. I usually hold a piece of cardboard or sheet vinyl against the glass as protection while nailing the inside molding. Another trick is to angle the nails slightly towards the glass. This angling gives more room for nailing and helps compress the caulk against the glass. Using pilot holes and screws would be another alternative. When the top and bottom moldings have been fastened, the temporary scraps which were used as stops can be removed and the side molding installed.

The trapezoidal windows on the north side were much smaller than the full-size windows, and thus easier to handle. It was convenient to work from the outside, so the inside molding was fastened first (Illus. 12–3). If you don't like handling and nailing the molding with all that sticky caulk

on it, there's an alternative. Install the windows *without* the caulk. Then, after all the windows are in place, go around with your caulking gun and place a small bead of caulk at both edges of the molding: against the window *and* against the frame. Do this on both sides of the window. It's very important to use top-quality caulk with this method.

PRE-HUNG UNITS

It's a fault of the original Log End Cave design that there aren't any windows that open. I trust the reader will remedy that situation with his own plans. Pre-hung window units, be they double-hung, casement, awning, whatever, are easy to install if you've left the right rough opening. Don't trust the figures given in the window catalogs. They give you all kinds of confusing dimensions: glazing area, unit size, rough opening.

Have the window unit on hand and make sure it fits the actual rough opening. The installation of the manufactured units can be accomplished in very much the same way as for the fixed units already described. Follow the manufacturer's installation instructions.

DOORS

You can buy pre-hung, insulated, and weather-stripped doors, or you can make and hang the door yourself. The first option is easier, and, if care is taken in the selection of a quality unit, you'll get a weather-tight door. I was sorely tempted to use factory-made doors, and we even spent an afternoon looking at various doors. Jaki convinced me that I could make a door more in keeping with the rough-hewn atmosphere that we were after. I took a day and built the door. (Make sure you have at least *two* doors!) It's 4″ thick, but who knows what it weighs?

I won't go into detail about our own door construction, as people who make their own doors like to do it their own way, but I do include a cutaway view of the door (Illus. 12–4). We stayed with this basic door design throughout all the buildings at Earthwood. In fact, I still refer to the original diagram to build new doors. The diagram has stood the test of time. Visitors are impressed with the thickness and mass of the door. Nowadays, I'm inclined to hang the doors with three barn hinges instead of two, so two doors can be hung with three pairs of hinges.

My nephew Steve and his friend Bruce came back up one weekend in November to see how things were going. We put them to work, of course, moving an old kitchen stove we'd bought at an auction and installing the homemade door with heavy barn-door hinges.

INSTALLING THE VENTS

Ventilation was one of the unknowns we had to deal with at Log End Cave. We knew we had to have 'em, as there weren't any windows that opened. Where should they be positioned? What size? How do you build them? The subject wasn't covered in the few articles about underground housing I was able to track down in 1977, so I set out to design the most flexible system I could, with

vents that I figured would be large enough to handle any situation short of a skunk in the stovepipe. The floor plan in Illus. 12–5 shows the location of the six exterior vents (the seventh is the door): one in each bedroom, one in the sauna or utility room, one in the living room, and two in the kitchen-dining area. By opening and closing vents or the door, we were able to create any cross-draft we needed for the removal of kitchen odors, for heating and cooling, or simply to encourage a flow of fresh air.

Our exterior vents were homemade. They were designed to fit unobtrusively into our 10″-thick cordwood-masonry walls. Construction was of aluminum mosquito screen and leftover 2 × 6s and 2 × 4s (Illus. 12–6). The vents were built on a flat surface such as the concrete floor, stained, and nailed into place. Jaki made the vent covers by gluing a piece of 2″ Styrofoam® to a piece of plywood corresponding to the shape of the vent surrounds, some of which were trapezoidal. The foam is carefully shaped to fit the actual vent space. The cover can be held against the vent surround with a screw (or thumbscrew) and a receiving socket. The cover can be decorated with material, paint, varnished wine labels, whatever.

During the winter of 1978, we experienced no shortage of oxygen and didn't need to use the vents at all, so we kept them closed to conserve fuel. The stoves got their required air from the under-floor vents. In spring and summer, we used the vents primarily to regulate humidity, opening them on dry days, closing them on humid days.

CORDWOOD MASONRY

The defining design feature at Log End Cave is the use of *log ends*, the individual building blocks which make up that style of building known variously as cordwood masonry, stackwall construction, stovewood building, etc. (I wrote about this style of building in Sterling's *Complete Book of Cordwood Masonry Housebuilding*.) The wall derives exceptional thermal characteristics by virtue of the special double-mortar matrix, the inner mortar joint being separated from the outer by an insulated cavity. It's amazing how many low-cost underground houses make use of this weird

① SILICONE CAULKING AT ALL THRU JOINTS
② 2"x6" CHECKED OUT FOR WINDOW FRAMING
③ 15-LB FELT UNDER OUTSIDE SHEATHING
④ 2"x6" FRAMEWORK HELD WITH TOENAILS UNTIL
    1" SHEATHING IS FIXED

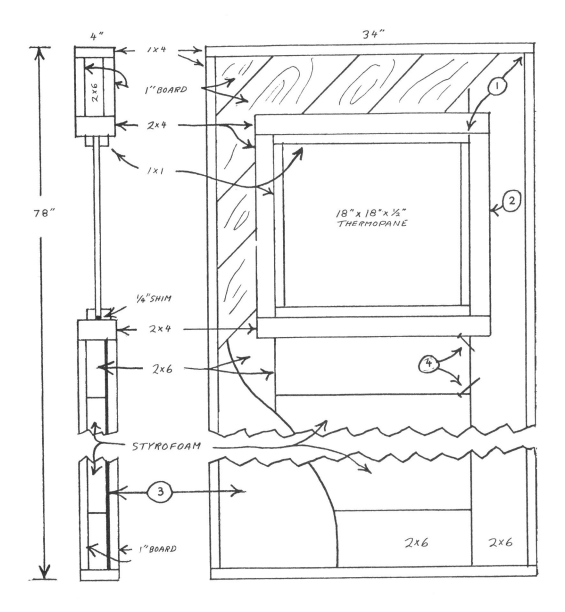

Illus. 12–4. This basic door design was used at Log End Cave and at Earthwood.

and wonderful building technique, as a glance through the case studies in this book will show. See the color section for some lovely examples of this technique. We feel that the cordwood walls contribute to the warm, rustic atmosphere we tried to create. And it's about the least expensive way I can think of to fill in a post-and-beam framework.

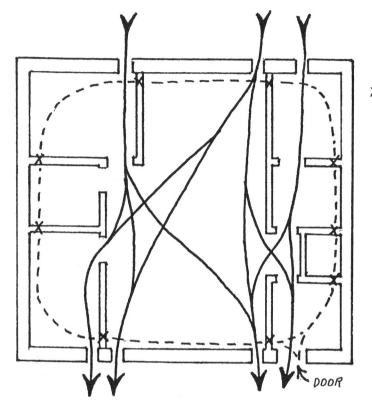

CROSS-VENTILATION PATTERNS

X = SUGGESTED LOCATION OF INTERIOR VENTS (AT FLOOR LEVEL)

DOOR

Illus. 12–5. Floor plan (left) showing the location of six exterior vents: one in each bedroom, one in the sauna (or utility room), one in the living room, and two in the kitchen-dining room. The door functions as the seventh vent.

Illus. 12–6. Vents (below) can be constructed from aluminum screen, plywood, weatherstripping, rigid foam, and scrap 2 × 4s and 2 × 6s.

VENT CONSTRUCTION

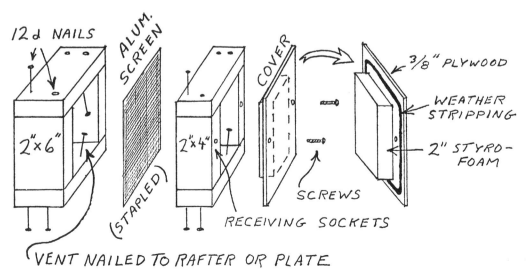

12 d NAILS

ALUM. SCREEN

2"×6"

(STAPLED)

COVER

2"×4"

SCREWS

RECEIVING SOCKETS

3/8" PLYWOOD

WEATHER STRIPPING

2" STYRO-FOAM

VENT NAILED TO RAFTER OR PLATE

13

Retaining Walls & Landscaping

STONE RETAINING WALLS

Because we were in a hurry to get the roof on, we backfilled the home before building the retaining walls on the south elevation. Actually, backfilling and building the retaining walls are very closely related; part of the function of the wall is to keep the backfill from spilling around to the front of the house. We should have built the retaining walls first. As it was, some of the roof soil had to be retained with 2×6 planks until the walls were complete.

The main cause of retaining-wall failure in northern climates is frost heaving, so good drainage behind the wall is imperative. For that reason, we continued the home's footing drains behind the large retaining wall and backfilled the wall with sand. The two small retaining walls would stand on their own without extra precautions, since they were little more than a single course of massive boulders. Even if frost was to shift them a little, there'd be nothing to come toppling down.

We completed the drains before starting the actual stonework. The footing drains meet at the southeast corner of the house (Illus. 10–4) and continue as retaining-wall drains until this large wall becomes a single course of boulders. At this point the drain becomes nonperforated 4″ flexible tubing, crosses the wide path that leads to the south-side entrance, and continues on to the soakaway (Illus. 10–3). The underfloor drains and the kitchen-sink drain (greywater system) take a separate path to the same soakaway. The soakaway is a 10′-diameter by 4′-deep hole filled with fieldstones and covered with 15-pound roofing felt and earth. It's better if the footing drains finally come out above grade, if your site allows, as there's always the possibility of the soakaway being full. As with septic-system drain fields, it's always a good idea to slightly crown the soil above a soakaway so that it doesn't become a depository for ordinary surface runoff.

The retaining walls are built of massive stones unearthed during the original excavation. We ended up with an incredibly strong and beautiful wall, while getting rid of a lot of pesky boulders that would have made landscaping difficult. Really nice stones, of course, can be used to help create a beautiful natural landscape.

A typical stone in the wall weighs several hundred pounds, so there's little alternative to building the wall with a backhoe. There must be good coordination between the builder and the equipment operator on this sort of work. Before the operator arrives, spend an hour or so looking over the stones in the heaps. Then have the operator push all the good stones to a central depot near the work. This centralization saves running around for the right stone when you need it.

The procedure is much the same as building any dry stone wall. Have the backhoe clear a flat base for the wall, depressed a couple of inches to give the large stones a bit of a bed in which to rest. Then, using a heavy chain with chain hooks at each end, the backhoe can lift the required stone and set it into position. First use the stones with two parallel faces for the bottom courses, saving stones with

Illus. 13–1. The large triangular spaces left between the boulders can be filled with smaller stones.

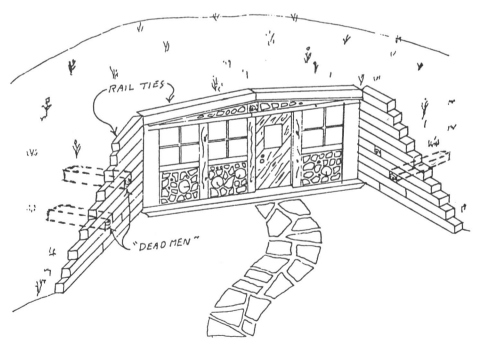

Illus. 13–2. Railroad ties or 6″ × 6″ landscaping timbers can be used to build retaining walls. The "dead men" are timbers set perpendicularly into the bank, and they help resist lateral pressure on the walls.

only one face for the top course. It will be necessary to shim with smaller flat stones during building, so keep a pile of these handy. After the backhoe has finished, holes can be filled with smaller stones (Illus. 13–1).

OTHER KINDS OF RETAINING WALLS

Not every excavation will yield megalithic stones. Retaining walls can be built with smaller stones, blocks, bricks, old railway ties, and landscaping timbers. I've even seen old tires used, although I wasn't impressed with their appearance. Whatever the material chosen, the principles are the same. Build on undisturbed or compacted earth, or, better, a shallow (3″-deep) trench filled with crushed stone. Backfill with porous material. Use perforated drain tile behind a high wall to carry away water.

When using railroad ties or landscaping timbers, include a few *dead men* in the wall. Dead men are 4′-long pieces of the retaining timbers placed at right angles to the face of the wall, and laid back into the earth (Illus. 13–2) during backfilling.

LANDSCAPING

The area in front of the exposed cordwood wall on the south side must have good positive drainage away from the building. Snow can collect there and snow always melts, even where I live. The frost-wall footing should be exposed 6″ above grade, and protected from rapid heat transfer with extruded polystyrene on the exterior. The ground in front of the frost wall/doorstep thus created should slope away from the building with a pitch of at least 1″ every three feet. Use a cement coating to protect any exposed polystyrene insulation from sunlight. Surface-bonding cement works great, if you have a little left over, or you can buy pre-mixed trowel-on products made for the purpose.

The area in front of the door was wide enough so that we could back a car or truck right up to the door. After bringing in groceries or firewood, we'd park the vehicles in a turnaround area away from the house. A good pathway up to the house would have 6-mil plastic on the exposed earth, large flat stones or 2″-thick concrete paving tiles on the plastic to create a walkway, and 2″ to 3″ of crushed stone between the paving stones. The

Illus. 13–3. The intent of the Log End Cave design was to have a minimal impact upon the landform.

plastic stops weed growth in this area. If you prefer green grass (and tight mowing) to bare stone, the area could be planted with grass. A brick or stone pathway could run down the center. This kind of landscaping decision is largely a matter of personal taste.

All too often, people don't give sufficient thought to the final appearance of the property. Remember that one of the best reasons to build underground in the first place is to create a habitat which harmonizes with its surroundings. An extra measure of grace is added to life when you can step back from your work and feel good about it, so don't put off the exterior-finish considerations any longer than necessary.

At Log End Cave there were two huge piles of earth to be spread out over the site, the larger on the east side. A bulldozer is the best machine to use to spread out the earth. Because we'd taken our east-side retaining wall all the way to the old stone wall which surrounds the meadow, we had an ideal depository for excess earth. On the west side, there was just the right amount of earth to achieve natural-looking contours away from the roof. The desired effect is that the roofline looks as little like an unnatural protuberance as possible. We felt that this effect was successfully achieved at the

Cave, and others have commented favorably on the way that the home seems to fit unobtrusively into its surroundings (Illus. 13–3).

Because of the lateness of the season, we postponed spreading the sods until spring. Then, after the sod roof was in place (Illus. 13–4), we spread seven loads of topsoil to finish the landscaping. We planted timothy and rye, then rolled and mulched the entire area. The grass came in well.

To keep down mosquitoes and blackflies in the spring and summer, we kept the area around the house mowed as far as the old stone wall that bordered the meadow. We built our first megalithic stone circle on the lawn just to the east of the home, using the last eight stones left over from the original excavation. Years later, at Earthwood, we refined our stone-circle technique to include standing stones of up to 7' high and weighing two tons. The experience of building the retaining walls at Log End served us well at Earthwood, which also featured megalithic retaining walls.

We left the forest wild beyond the old stone wall, although we did remove a few small coniferous trees which blocked sunlight from the home in winter, when we really needed the solar gain. Removing these trees gave us the added benefit of being able to look further into the forest.

Illus. 13–4. The sod roof is in place. The north gable windows admit light to the back rooms.

14

The Interior

NOTHING SPECIAL

There's nothing special about building internal walls in an underground house. Simple stick framing with 2 × 4 studs actually works very well. Such a frame will receive panelling, gypsum board, rough or finished boarding, almost any surface. We did most of the framing before backfilling the home so that we could help buttress the walls. It's easier to build the frame for a panel on the floor and then stand it up (Illus. 14–1) than it is to toenail each stud to the floor and ceiling plates.

Because finished 2 × 4s actually measure 1½″ × 3½″, we made the floor plate fast to the floor with 2″ masonry nails. Twelve-penny (12d) commons are fine for all other framing purposes. As

discussed previously, our interior was designed so that we only had to fit one wall around rafters. For the most part, we used ½″ drywall (*gypsum board* or *plasterboard*) for the interior walls, painted with white-textured paint. We make our own textured paint at great savings by mixing one part cheap white latex paint with five of pre-mixed joint compound. Mix the ingredients thoroughly with a stick and apply with a roller. You can buy textured rollers or make your own by scorching an ordinary roller with a propane torch. Such a roller gives a nicely textured 3-D look to the wall, although you'll use quite a bit of the mixture.

FILLING IN BETWEEN RAFTERS

Our rafters are exposed and carried by three large barn beams, and the internal walls are planned to meet the underside of the beams. In order to diminish sound transfer from the living to sleeping or bathroom areas, we had to fill the space between rafters (from the top of the beam to the roof planking). But how? A simple and practical solution, well in keeping with the motif of Log End Cave, was to use cordwood masonry. For these internal areas, we used 5″ log ends and laid them with a full-width mortar bed. Unlike external cordwood walls, there's no need to include an insulated space between the inner and outer mortar joints.

The mortar mix for cordwood masonry is 9 parts sand, 3 parts sawdust, 2 parts Portland cement, and 3 parts builder's (or *Type S*) lime, equal parts by volume. Use sand that's sugary in texture rather than coarse, as the mix will be less crumbly.

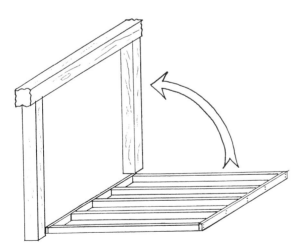

Illus. 14–1. When fashioning interior walls, it's easier to build the frame for a panel on the floor and then stand it up, rather than toenail each stud to the floor and ceiling plates.

Illus. 14–2. Log ends fill the space between rafters from the top of the beam to the roof planking. They go well with the rough-hewn style of the house.

The sawdust should be softwood, such as pine or fir, passed through a ½″ screen, and soaked overnight to absorb water before it's added to the mix. The use of dry sawdust would be a big mistake. It would rob moisture from the mortar, accelerating the mortar set (curing time), instead of retaining moisture and retarding the set. The mortar is left recessed so that the wood is "proud" of the masonry background by ¼″ to ½″. Finally, the mortar is smoothed with a pointing knife made from an ordinary stainless-steel butter knife which has had its last inch bent to a 30° angle.

Each panel between rafters has some different design feature, such as symmetry, a specific pattern, or a particular kind of rubble effect. Cordwood masonry (Illus. 14–2) ties the interior architecturally to the exterior, and provides an interesting and pleasing design feature to the open-plan living-kitchen-dining area. Also, it's fun to do.

Alternatives to cordwood masonry in these locations would be: mortared bricks, sections of 4 × 8 timbers, or drywall sections nailed to furring strips.

HEAT SINK

The room divider between the living and dining areas is of solid stone masonry a foot thick and, as both the kitchen and parlor stoves had their backs to the stonework, the room divider acts as a *heat sink* or *storage heater*. A mass of stone 5′ high, 8′ long, and 1′ thick, it has forty cubic feet of stone masonry. At an average weight of about 145 pounds per cubic foot, the mass weighs 5800 pounds, almost 3 tons. That's a lot of BTUs of heat-storage capacity, and it'll store "coolth" in the summer. Because the stone mass isn't in direct contact with the earth (remember the under-floor insulation), all the stored heat from the stoves is given back into the room as the interior temperature tries to fall below the mass temperature. The heat stored in the mass exerts a moderating influence on the room temperature. The block walls and the floor also act as heat sinks, but they don't get "charged up" quite as well as the stone heat sink located right next to the radiant stoves.

Building the room divider was fun, although I was hesitant about starting because I'd never built a freestanding stone wall before. Building the wall and a large stone hearth on the floor took 5½ days, with Jaki doing all the mortar pointing with the homemade pointing knife. We took the time to include several design features, like stone shelves and a massive stone table (Illus. 14–3), stones with fossils, and a raised seat in the hearth in the shape of a butterfly (Illus. 14–4). My mortar mix was 5 parts sand, 1 part Portland cement, and 1 part masonry cement. It's a strong mix, light in color, recommended to me by a mason. The stones were taken from the old stone wall (in front of the house) which had to be cut through with the bulldozer to provide vehicular access to the front door.

At Earthwood, we took the heat-sink idea to another level. There, we included a 23-ton *masonry stove* at the center of the home. The masonry stove has a firebrick-lined firebox at its heart. We actually charge this great mass from within. Wood is deliberately burned hot and fast, so that there are no unburned wood gases to be wasted up the chimney, or, worse, to condense *in* the chimney in the form of dangerously flammable creosote. The heat from the hot burn is transferred to the stone (or brick) by way of a series of horizontal flues. When the charge of wood is reduced to a bed of red-hot coals, with no more colored flame, the stove is closed down almost airtight. A *blast gate* (damper) is closed so that the coals last a long time, continually charging the mass. Such a hot

Illustrations 14—3 and 14—4. Left: The room divider includes stone shelves and a massive stone table. Above: The hearth features a raised seat in the shape of a butterfly.

burn is the most efficient way to burn wood, about 35% more efficient than using most wood stoves, and the house (even a surface house) stays at a stable temperature thanks to the BTUs stored in the 23-ton mass (Illus. 14—5).

WIRING & PLUMBING

Because the floor level of the underground home is normally several feet below that of a surface dwelling, special care must be taken to assure that the septic system is downgrade from the home and that the house drain has a positive slope. Outside of that, plumbing and wiring are really no different from the systems used in any other house.

At the Cave, and again at Earthwood, we made use of alternate systems, solar (*photovoltaic cells*) and wind energy. Neither home is connected to commercial power lines. At Earthwood, we pump all our water with a piston pump married to an old bicycle. Six or eight minutes of pedalling a day provides all the water we need for our family of four.

Illus. 14—5. The masonry stove at Earthwood

Electrical wiring can be included in the internal frame walls, in the same way as in conventional houses. Because the floor is concrete, the National Electrical Code (NEC) requires that all plugs, called *duplex receptacles* (DRs), must be ground-fault protected, either by the use of *ground-fault-interrupted duplex receptacles* or by the use of ground-fault circuit breakers in the electrical-supply panel box. With ground-fault-protected devices, the danger of electrical shock is eliminated, particularly in highly conductive locations, such as near a sink or where a concrete floor might get wet. The ground-fault interrupters sense a conduction of electricity to a location other than where it's intended to go (the ground fault), and the GFIs trip instantly to an open position. This happens so fast that the person who might provide a ground won't even feel a shock. Years ago in Scotland, Jaki was filling an electric kettle from the kitchen sink while the kettle was still plugged in. She experienced 240 volts (standard house current in the United Kingdom), which she described as like being hit hard in the back with a plank. Had the circuit been ground-fault protected, she would have missed this unique experience.

It will probably be necessary to have at least a few duplex receptacles around the perimeter of the floor plan, in order to satisfy the code, which requires a duplex receptacle at least every 12' along a wall. With concrete or concrete-block walls, and, incidentally, with cordwood masonry, the easiest way to install DRs is by using either metal conduit or code-approved surface-mounted wire mold. Wire-mold systems are made by several companies, and come in different styles and colors. Wire mold is often used now in commercial buildings, even restaurants, which might have stone or brick walls. It's attractive, safe, and easy to alter should a change be desired. Another advantage is that use of conduit or wire mold allows you to build the basic structure first and wire it later. One manufacturer I've used is the Wiremold Company. Ask them for their catalog and wiring guide.

Similarly, plumbing can be located within internal framed walls. And, personally, I'm not disturbed by an occasional pipe which has to be left exposed, although the reader may feel differently. Exposed plumbing can be repaired or added onto easily.

FLOOR COVERING

The floors at the Cave were covered with sheet vinyl in the kitchen and bathroom, and carpeting in all other rooms except the mudroom and the utility room, where concrete floor paint was used. Because of our homesteading life-style and two German shepherds with perpetually muddy paws, we chose a used industrial-grade (short nap) carpet in almost-new condition. We carpeted half the house for $120 plus a little for a foam pad, important for extended life of the carpet. Friends in the carpet-laying trade laid both the carpet and the sheet vinyl in exchange for seven weeks' free rent at the Cave while we visited Scotland, a good deal all around.

It's possible to construct a wooden floor on furring strips laid over the concrete floor, but this reduces the value of the concrete floor as thermal mass. If a wooden floor is desired, you could leave out the concrete floor altogether. Cover the earth with 6-mil plastic and run floor joists from footing to footing, keeping the joists a few inches clear of the plastic. Vent the underfloor space with floor registers. You might even consider running ductwork under there if a forced-hot-air system is planned. A warm floor translates into a warm house. A *grade beam* (a fancy term for an internal footing) or two might be necessary to reduce floor-joist spans. Another benefit to this type of floor is that the underfloor space can be used for running plumbing and electrical wiring.

Another flooring alternative is to apply hardwood parquet right to the concrete. This type of floor is applied in tiles or in lengths of hardwood flooring. Both systems involve the use of a mastic made for the purpose.

In the past few years, I've become enamored of floors made of old recycled roofing slates bedded in a thin mortar joint right over the concrete floor. With a brush or roller we apply a coat of bonding agent to both the concrete floor and to the underside of the slates to be installed. The "underside," in this case, is actually the side that used to be exposed on the slate roof.

Using wooden lath as a screeding guide, we lay down a ⅜" bed of strong mortar (5 parts sand, 1 part Portland cement, 1 part masonry cement). Then, with a rubber mallet, we set the slates into the mortar, trying to keep them all in the same

Illus. 14–6. Point the mortar between slates with a pointing tool made from a steel butter knife.

Illus. 14–7. Jaki, at Earthwood, seals the floor, which is made of roofing slates set in strong mortar.

plane. Finally, any excess mortar is scraped away with a trowel, and the 1″ mortar joints between slates are pointed with our old friend, the home-made pointing knife (Illus. 14–6). When the mortar is fully cured, in about four days, a slate sealer can be applied to seal the floor and bring out the interesting colors of the slate. We initially apply two coats of sealer, and we reseal the floor every year or two (Illus. 14–7).

Malcolm Wells likes a particular concrete-floor technique that he calls the least expensive, the toughest, the best for solar-energy absorption, and the least slippery of all finishes. Wells says:

The final product looks for all the world like antique Italian leather, and we've fooled many people into believing that that's exactly what it is when they rave about it.

In short, Wells screeds the concrete floor, but doesn't trowel it smooth. After the construction work is finished and the house is ready to be occupied, he applies a "cheap sealer" and then two coats of a mix of 3 parts urethane floor sealer and 1 part wood stain, using a long-handled roller.

The liquid mixture is, of course, deepest in the little valleys between the tiny ridges, so it remains darkest there. The overall color is a dark antique brown. The best color to use . . . is Minwax's [stain] called "Jacobean."[8]

Concrete floors are hard on the skeletal system, as two hours' shopping in a mall will quickly confirm. Slate, sheet vinyl, hardwood tile—even Wells' *faux* leather—don't help much. Consider using throw rugs, rush mats, or padded carpet where you'll be doing a lot of standing; or use a wooden floor.

Furnishing and decorating are matters of individual taste, of course, and I've really no intelligent comment to make except to advise keeping the walls and floors as bright as possible. The exception would be a floor which receives direct solar gain. A dark floor absorbs heat better than a light-reflecting floor.

15
Performance

We lived in Log End Cave, summer and winter, for three years. We've seen the outside temperature vary from −40° to 90°F, while the interior temperature ranged from 58° to 77°F. The drainage and waterproofing systems met the severest possible test one spring when it rained steadily for three days on top of 36″ of compressed snow. The only problem encountered during that time was one or two minor flashing leaks at the base of the front windows, easily repaired.

Other problems have occurred, and I'll relate them now with comments about correcting them.

ENERGY "NOSEBLEED"

On the south side of the home, at the corners, the concrete-block walls transfer heat from the interior to the exterior. Or, think of "coolth" being transferred from the exterior to the interior. It's all the same to me: energy "nosebleed." We didn't notice it until December of our first year in the Cave. The bedroom wall, and, later, the mudroom wall, were getting wet at their southern ends. Our first thought was "Leak!" After all, the waterproofing detail at those corners was particularly tricky. Luckily, before tearing up all that earth, I remembered something I'd read by Wells about conduction of heat at parapet walls and the need for a thermal break.

I quickly covered the exterior of the block wall with 2″ of beadboard. Within 48 hours, the wall had dried up, its temperature now able to rise above dew point. This sort of detail should be considered at the design stage. One solution is to install the rigid foam during construction, held in place by the retaining wall or mechanical fasteners. Then cover the foam with a protective coating to prevent UV damage.

LEAKS IN SKYLIGHTS

Proper flashing of skylights is tricky business, and we had flashing leaks early on. Back in chapter 11, I described the flashing process as it should be done. In 1978, we weren't as careful in our detailing, and didn't know about the Bituthene® membrane, which would have made waterproofing easier and better than our efforts with the black plastic. But our main error was in not providing positive drainage around each skylight. Leakage in a roof isn't likely to occur on the wide-open spaces, where the membrane can be easily applied. Leakages are far more likely to occur where there's any projection (such as a chimney or skylight) through the roof.

We cured our flashing leaks by digging up all the earth within 12″ of the skylights and the main

Illus. 15–1. Skylights, chimneys, and other roof projections should be surrounded by good drainage material, such as #2 crushed stone.

stovepipe. We then cleaned and dried the area thoroughly and applied plenty of roofing cement and 6-mil black polyethylene, well lapped. Finally, we used sand to backfill the skylights and spread wood chips over the sand for the sake of appearance (Illus. 15–1). The sandy areas around projections were connected to each other with 4″ nonperforated flexible tubing. The bathroom and office skylight areas were drained by the same tubing to the sandy backfill at the edge of the building. The 4″ tubing was sod-covered and invisible. Nowadays I use crushed stone instead of sand as drainage around any projections through the roof. Stone drains much faster than sand, preventing any chance of a pool of water collecting next to flashing details.

PRESSURE ON RETAINING BOARD

The earth roof exerted much pressure on the 2 × 12 retaining boards, probably because of expansion of the wet or frozen soil during the winter. By May the boards were almost 1″ out of plumb. The cure wasn't difficult. We removed the topsoil within 5″ of the retaining boards, and replaced it with a thin layer of sand over the membrane and a 6″ depth of #2 crushed stone up to the surface

Illus. 15–2. Removing the sod within 5″ of the retaining board and replacing it with a layer of sand covered by 6″ of crushed stone improved the drainage and relieved the lateral pressure from the retaining board.

(Illus. 15–2). Thus the space next to the retaining boards was now well drained, removing the hydrostatic pressure.

CONDENSATION

In late spring we began to notice condensation around the base of the wall. Here's what happens: Because of the so-called "flywheel effect," the earth's temperature is still quite cool at 6′ of depth in May and June. This "coolth" is conducted up through the footing to the area where the poured floor meets the first course of blocks. Warm moist air hits the area, followed by: dew point, condensation, damp. Because we'd failed to insulate right around the footings and under the floor, we didn't have a lot of choices by way of cure. Someone connected to commercial electricity could have run a dehumidifier. We didn't have that option. We vented in the corners where internal walls met the external walls, as per the dotted line shown in Illus. 12–5. We rolled the carpet back from the wall, so that the rug wouldn't get mildewed. We used some moisture-absorbent material to dry things out. We toughed it out, remembering architect John Barnard's warning that underground houses have a high humidity until all the concrete fully cures.

Don't get nervous. Whenever possible, leave the building open on dry days. Drying out may take two years.[10]

In our case, we also had a good ton of water trapped in the 4 × 8 hemlock rafters, which had been trees just days before they were installed. Hemlock has what is known as 110% moisture content, which means that the green wood weighs 110% more than it does when the wood is fully dry. Because of the waterproofing, all that green-wood moisture can only come into the house. Wood takes about a year per inch to dry through side grain. Using woodstoves helped dry things out quite well, and we used the stoves every two or three days during that first June.

The second spring was better. Humidity was lower, but there was still some condensation. The third year was even better, but not completely free of condensation in the corner rooms. An ounce of prevention is worth a pound of cure, as we found at Earthwood, where we wrapped the footings with insulation as described in this book. We've had no wall or floor condensation at Earthwood. Humidity is comfortable, upstairs and down.

LIVABILITY

The *livability* (the quality of life in the home) of Log End Cave exceeded our expectations. There was plenty of light; heating and cooling characteristics were phenomenal; the view to the south was excellent. The home was rustic and cozy. There were unexpected joys, too, like the sun's rays on summer evenings giving natural illumination to the dartboard, and being able to lie back in the bathtub bathed by sunshine from above. Of course, it's possible for folks to walk onto the roof and peer into the skylights, but they're unlikely to do this more than once.

Our goals from the outset were: to create a

Table 3
Amount of Other Fuels Equivalent to a Cord of Air-Dry Wood

A Cord of Air-Dry Wood equals		Tons of Coal	Gallons of Fuel Oil	Therms of Natural Gas	Kilowatt Hours of Electricity
Hickory, Hop Hornbeam (Ironwood), Black Locust, White Oak, Apple	=	0.9	146	174	3800
Beech, Sugar Maple, Red Oak, Yellow Birch, White Ash	=	0.8	133	160	3500
*Gray and Paper Birch, Black Walnut, Black Cherry, Red Maple, Tamarack (Larch), Pitch Pine	=	0.7	114	136	3000
American Elm, Black and Green Ash, Sweet Gum, Silver and Bigleaf Maple, Red Cedar, Red Pine	=	0.6	103	123	2700
Poplar, Cottonwood, Black Willow, Aspen, Butternut, Hemlock, Spruce	=	0.5	86	102	2200
Basswood, White Pine, Balsam Fir, White Cedar	=	0.4	73	87	1900

Assumptions—

Wood: 1 cord = 128 cubic feet wood and air or 80 cubic feet of solid wood at 20% moisture content. Net or low heating value of one pound of dry wood is 7,950 BTU. Efficiency of the burning unit is 50%.

Coal: Heating value is 12,500 BTU per pound. Efficiency of the burning unit is 60%.

Fuel Oil: Heating value is 138,000 BTU per gallon burned at an efficiency of 65%.

Natural gas: One therm = 100,000 BTU = 100 cu. ft. Efficiency of burning is 75%.

Electricity: One KWH = 3,412 BTU. Efficiency is 100%.

*This is the quality of wood our tests are based on.

home with superior heating characteristics; to own it ourselves instead of the bank owning it for us; and to maintain the beauty and natural harmony of the land.

HEATING

Over three years, our average fuel consumption at the Cave was between 3 and 3¼ cords of medium-grade hardwood per year. This converts to about 360 gallons of fuel oil or about 9500 kilowatt hours of electricity (see Table 3). Earthwood uses 3¼ to 3½ cords of wood per year, but has over twice the usable floor area as the Cave, so it's twice as efficient in terms of heating units per square foot of living space. I attribute this efficiency to Earthwood's round shape (the least skin to enclose a given area), insulation better than what we'd used previously, 16″ cordwood walls above grade, and the use of a highly efficient masonry stove. There's

little doubt that a masonry stove at Log End Cave would have reduced wood consumption to 2½ cords or less. As I write in 1994, we can still buy good hardwood for less than $100 per four-foot cord, so we're talking about an annual heating cost of $300 to $350 per year, even if we have to buy our firewood.

While living at the Cave we found that we burned 95% of our fuel in our airtight kitchen cookstove. We fired the parlor stove fewer than ten times a year, occasionally to charge the thermal mass of the north wall during particularly cold times, and sometimes just to enjoy the stove's open fire.

We found that the home required less fuel when there was a good snow load on the roof. During the winter of 1979–80, although considerably warmer than the record-breaking cold winters of 1977–78 and 1978–79, we actually required a little more fuel because of the total lack of snow until

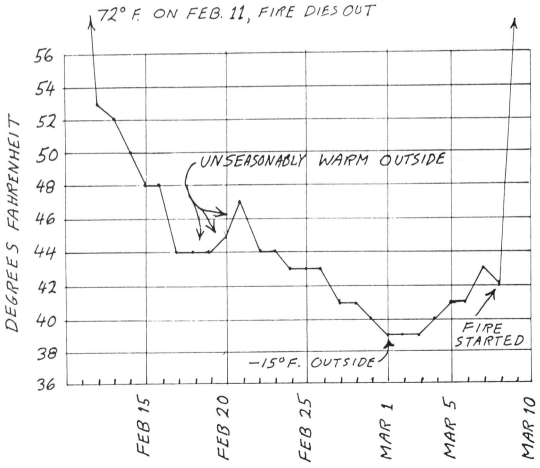

Illus. 15–3. Winter without heat at Log End Cave

mid-February and the resulting frozen roof. On February 11, 1980, we left for a month's vacation. A friend living next door at Log End Cottage came in to the Cave every day to look after the dogs and to record the home's inside temperatures. There was no heat in the Cave except for solar gain and the heat produced by two German shepherds. Windows and skylights weren't shuttered. The temperature in the house bottomed out at 39°F on March 1 (with the temperature −15°F outside) and then began to rise (Illus. 15–3). It should be noted that the temperature was always taken at 7:30 A.M., the very coldest time of day in the house. On sunny days, the internal temperature would rise six or eight degrees by 2 P.M.

Two worthwhile points are shown by Illus. 15–3. One is that earth-sheltered homes, set in the above-freezing earth temperature, can be left for extended periods in the winter without employing special efforts (such as threshold heating) to keep pipes and food from freezing. Illus. 15–3 also reflects the primary reason that an earth-sheltered house is easy to heat: The home is set into an ambient temperature that's warmer than the outside air. This theory was presented way back in chapter 2 (Illus. 2–8).

COOLING

In addition to the benefits of evaporative cooling off the earth roof, there are two more reasons for ease of cooling in underground houses. One is the "thermal flywheel effect." The earth's coolest temperature lags about two months behind the climate's lowest average temperature readings at a depth of 5.3 feet. The earth just outside the walls is likely to be coolest, then, in early March, gradually warming up to its warmest temperature in early September. (Another great thermal mass, Lake Champlain, performs similarly: The highest water temperature is generally recorded about September 1.) This thermal lag also helps with heating during the first months of the heating season.

The second cooling advantage is that the underground house, with its characteristic slow temperature change (thanks to the moderating effect of its own great mass as well as the surrounding cool earth), can effectively average out variations in temperature over a day, a week, or even a

month. In an underground home, temperature curves have shallower slopes than those of surface homes.

We've experienced these cooling advantages at both the Cave and again at Earthwood. Others report similar observations. Underground houses don't require air conditioning, although they may require some form of dehumidification at certain times of the year, particularly in the southern U.S.

ECONOMY OF CONSTRUCTION

The methods of construction described in this book have been chosen because they can be readily learned by the owner-builder, and they're moderate in cost compared with poured-concrete walls, pre-stressed concrete planks, and the like.

The materials and contracting (excavation and landscaping) for Log End Cave cost $6750.57. We also spent $660 on labor, bringing the total cost of the basic house to $7410.57. Covering the floors with good used carpets and new vinyl (concrete paint in the mudroom and utility room) added $309.89, and fixtures and appliances added another $507, bringing the total amount spent on our house to $8227.46. I'm sure that this figure is within 3% of the actual spending, as we kept receipts for practically every item. An itemized accounting appears in Table 4. The final total doesn't include the cost of our septic system, well, wind plant, storage batteries, or furniture, all of which were already in use at the Cottage. The cost of the septic tank and the drain field in 1975 was about $600. In 1994, a similar system with a thousand-gallon concrete tank and a medium-size drain field would (if contracted out) run about $2200 in our area. A lot of money can be saved if you can operate equipment yourself. The wind plant, tower, and batteries cost about $800. We later upgraded to a better wind plant. The spring-fed well was already there when we purchased the property.

These figures don't place a dollar value on the owner-builder's labor, but I believe this value should be measured in terms of time, not money. Money saved is a lot more valuable than money earned, because we have to earn so darned much of it to save so precious little.

Granted, the cost figures are old. People ask me what they could expect to spend for such a home in

Table 4
Log End Cave—Cost Analysis

Heavy-equipment contracting	$892.00
Concrete	873.68
Surface bonding	349.32
Concrete blocks	514.26
Cement	47.79
Hemlock	345.00
Milling and planing	240.26
Barn beams	123.00
Other wood	167.56
Gypsum board	72.00
Particleboard	182.20
Nails	62.88
Sand and crushed stone	148.21
Topsoil	295.00
Hay, grass seed, fertilizer	43.50
Plumbing parts	124.95
Various drainpipes	166.23
Water pipe	61.59
Metalbestos® stovepipe	184.45
Rigid-foam insulation	254.93
Roofing cement	293.83
Six-mil black polyethylene	64.20
Flashing	27.56
Skylights	361.13
Thermopane® windows	322.50
Interior doors and hardware	162.80
Tools, tool repair, tool rental	159.43
Miscellaneous	210.31
Materials and contracting cost of house, landscaping, and drainage	$6750.57
Labor	660.00
Cost of basic house	$7410.57
Floor covering (carpets, vinyl, etc.)	309.89
Fixtures and appliances	507.00
Total spending at Log End Cave	$8227.46

1994. Since 1977, concrete, concrete blocks, wood products, and heavy equipment have all just about doubled in price in our area. Improved techniques and building materials described in this book will add another 20% or so to the cost. My best guess is that a Log End Cave type of home can still be built in rural areas for a materials cost of about $16 per square foot, or $16,000 for a 1000 sq. ft. home, not counting lot, well, septic, and gold doorknobs.

Roughly 1800 hours of work were required to build Log End Cave. This includes about 450 hours of outside help, some of it paid, some of it volunteered. The bulk of the work was done during a five-month period ending on December 17, 1977. During January and February 1978, we completed most of the interior work. Final landscaping and placement of the sod was done in June 1978. For about half of the days, I noted actual hours worked in my diary, especially when this was required to pay help, but for the other half I've had to read my notes to see what was done and guess at the labor time. Therefore, the 1800-hour figure is probably not as accurate as the cost analysis, but I feel sure that it's within 200 hours of reality. Each builder will work at a different rate anyway, so the actual time it takes someone else will be different.

NOW HEAR THIS!

More than a third of the average American's after-tax income is devoted to shelter, usually rent or mortgage payments. If a man works from age 20 to age 65, it can be fairly argued that he's put in 15 years (20 in California!) just to keep a roof over his head. With a piece of land, six months' work, and, say, $16,000, he and his family could have built their own home.

To save 14½ years of work, you can't afford *not* to build, *even if it means losing a job*. Granted, the land and the $16,000 has to come from somewhere, but this amount is no more (and probably less) than the down payment on a contractor-built home, and about half the cost of a new mobile home (figuring either to be the same square footage as an underground home). The sum is about what many people pay for their *car*, which depreciates rapidly. And you can't compare what you get for your bucks.

What *do* you get for your time and money? You get a comfortable, long-lasting, energy-efficient, environmentally compatible, low-maintenance home. You get the design features that suit you, so that the house fits like an old slipper. You get built-in fire, earthquake, and tornado insurance. You get intimate knowledge of the home so that when maintenance or repairs are required you're the one who's best placed to make them. You get tremendous personal satisfaction; and you get freedom from a lifetime of economic servitude.

16

Case Studies in the North

EARTHWOOD

West Chazy, N.Y.

We were comfortable at the Cave, but the family was growing and the building wasn't designed for expansion. Underground houses are difficult to add on to unless expansion is addressed specifically at the design stage. A few years ago, a young couple with several small children visited us to discuss earth-sheltered housing. They insisted that they absolutely *had* to have 1600 sq. ft. of living space, *minimum*. The trouble was that they could only afford 800 sq. ft. of living space. I advised them to build the 800 sq. ft. that they could afford, leaving the east wall of surface-bonded blocks insulated, but not backfilled. As they were both making good money, they could afford to complete the other 800 sq. ft. two or three years down the road. This expansion could be accomplished by reusing the rigid foam (as long as it had been protected), and utilizing the internal wall as a thermal flywheel and effective noise buffer between one side of the house and the other (Illus. 16–1). An alternative detail would be a masonry stove in that east wall,

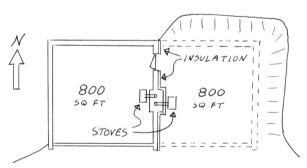

Illus. 16–1. Add-ons must be planned for at the design stage. The central masonry mass could incorporate a masonry stove.

later to be the central divider. The result: an energy-efficient, debt-free home. The trade-off: two or three years of less-than-desired living space.

The opportunity to build Earthwood came when, as the result of a contract gone awry, Jaki and I acquired some of the most important materials we needed to build a 39'-diameter round cordwood home, and a piece of land which we accepted back from a neighbor as satisfaction of a mortgage on a larger parcel we'd sold some years earlier.

The house site was a mess, two acres of badly scarred gravel pit. Taking a leaf from Malcolm Wells's extensive notebook, we decided to build on this marginal property and render it living and growing and productive again. We need to reverse the 20th-century attitude of building in all of nature's most beautiful places. Instead, let's reclaim blighted lands and make them beautiful.

We figured that it wouldn't add a lot to the cost of the home to build a second storey, since the price of the roof and foundation, two of the most expensive items, would be constant. Same with the well, septic system, and driveway. This time we'd build a house we could grow into, and that's indeed how things turned out.

Although Earthwood looks completely different from Log End Cave, it is, in many ways, an evolution of the same design, and makes use of many of the same techniques: post-and-beam, plank-and-beam, surface bonding, cordwood masonry. We corrected the glitches we'd encountered at the Cave, and retained the positive features, such as the large open-plan living-kitchen-dining area, and the positioning of the bedrooms away from the centrally located heating system, this time a masonry stove. Earthwood has twice as

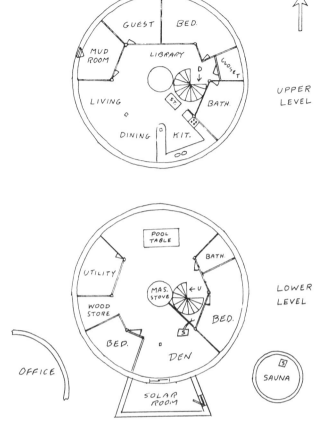

Illus. 16–2. Earthwood floor plan

Illus. 16–3. View of Earthwood and the Stone Circle. Photo was taken from halfway up the wind plant.

much space as the Cave, but the new house (Illustrations 16–2, 16–3) uses about the same amount of fuel as the Cave used, while keeping a very constant temperature.

RICHARD & LISA GUAY

Richard was one of the first students at our school, and *the* first, I think, to build from the 40' × 40' Cave plans. The home is located in Champlain, N.Y., just 3 or 4 miles from the Québec border. The basic plan (Illustrations 2–5 and 2–7), was modified only slightly to accommodate individual needs. Richard was single when he first built the home (Illus. 16–4). He married Lisa a while later, and the couple now have a daughter of six, Moni-

que. So the house was finished in stages, internally, between 1980 and 1986, always on a pay-as-you-go basis. Like most of the homes in this chapter, there were no bank loans, no mortgage.

When wood was the primary source of heat (he used a small amount of electric backup) Richard burned about 15 face cords a year, which translates to about 3¾ full cords, or 15% more wood than we used at the original Log End Cave. As his useful square footage was about 58% greater than the original Log End Cave, 1440 sq. ft. versus 910 sq. ft., the Guay home was very efficient, even with electric-backup heat figured in.

There have been two recent changes to the home, precipitated by Lisa's becoming a five-time champion on "Jeopardy." The couple spent a chunk of the winnings on a 9' × 40' solar room attached to the south side of the home. The addition, a manufactured unit, was professionally installed, with all low-E insulated glass. It supplies a wonderful break to our long North-Country winters and a super play area for Monique.

The other major change was the installation of oil-fired hot-water baseboard heat, which greatly simplifies heating the home. Their fireplace insert is still available for emergency backup, however. The small efficient oil furnace is located in the far corner of the new solar room (Illus. 16–5).

Illus. 16–4. Richard Guay's 40' × 40' Log End Cave during construction in 1980. The home is now fully earth-sheltered: berms on the north, east, and west sides; earth on the roof.

Illus. 16–5. The new oil furnace occupies a corner of the solar room.

We sometimes visit the Guays with a class of underground-housing students. During my last visit, I asked Richard what he liked, what he'd change, etc. He cited the low maintenance, no freezing pipes, good noise reduction, and harmony with the environment as the positive features. The house is much brighter than he expected it to be. Changes? He'd install in-floor heating right from the start and use 16"-thick cordwood walls instead of 12".

UNDERGROUND STUDIO

Elaine Cosgrove Rielly

Elaine has lived in and around Jay, N.Y., for most of her life. She was brought up just down the hill from her underground home in the farmhouse where her mother still lives. In fact, the property where she built her home, land with a wonderful view of the Adirondacks, was given to her by her parents.

Elaine and her husband started the home in July 1981 with a $5000 grubstake, and while they built, they lived in an old trailer. They moved into the 32'4" × 42'4" underground home in October of the same year, but it was "pretty rough," according to Elaine. They made things less rough over the years by paying for improvements as they could

afford them, thus avoiding debt. Elaine's father, who had a background in construction, helped the couple design a roof frame which would support the 6" earth roof, and then he helped to build it. One night a couple of years after the earth was placed on the roof, Elaine was startled to find that their 3 horses and 4 donkeys were all grazing happily on the lush roof grass.

Walls are 8" poured concrete, and the home is wrapped with Styrofoam® insulation, including 2" of the foam under the floor. Like so many builders of low-cost underground homes, the Cosgroves choose wood heat, and they incorporated at the center of the house a magnificent stone masonry chimney, measuring 5' square by 6' high, and tapering to 3' square for the next 6' before exiting the roof. Two wood stoves are vented into this mass, which weighs some 16½ tons. This beautiful stonework exerts a moderating influence over the temperature of the home, in both summer and winter. Eight face cords (2⅔ full cords) are used each winter to heat, and they have about the same climate as we do, close to 9000 degree days. As the useful living area is about 40% greater than the original Cave (1271 sq. ft., not counting a small sleeping loft, versus 910 sq. ft.) and it uses only 85% of the wood which we used, I think it's fair to say that the home is extremely energy-efficient. The main differences between the two homes are the 2" of Styrofoam® under the floor and the internal chimney's ability to store a lot more heat than the Cave's relatively small 5-ton mass.

Elaine is on her own now, and the home has become the Underground Studio, where she pursues her various artistic talents, including underground-design work using a CAD drawing program on her computer. Her 27'-long mural of the Adirondack countryside graces the living room and studio. Creative works and books on art are everywhere. The house is alive, inside and out. See the color section for photos of this house.

SIEGFRIED BLUM

Priceville, Ontario, Canada

"We built our earth-roofed round cordwood home in 1987, slaving nine long months from morning till night, using one of Rob's earlier books as a manual. However, we decided on a single-storey version, adding our own touches as we went along. Would we do it again? There's a loaded question, not easily answered. Yes, we like our unique home (Illustrations 16–6 and 16–7) and its special qualities, like the small amount of fuel required to keep it snug in winter. Our wood stove is normally lit around 4 P.M. and left to go out by midnight. The heat stored by the house keeps the temperature cozy until the next afternoon.

Part of our goal was a low-maintenance home, since retiring on a low, fixed income doesn't allow for lavish expenses such as painting the exterior, or reroofing. A lot of chores connected with the ownership of a conventional home are non-existent here.

The primary method of construction (cordwood masonry) lends itself well to owner-building. But beware, the patience required for seemingly endless, repetitive work, such as mixing 'mud' and laying up the log ends, becomes apparent only after you've started, and then there's no turning back! It took us seven weeks to complete the perimeter wall of our round house.

We decided to rough-plaster the inside of the cordwood walls, applying the same mix we used for the laying up of the log ends, a mix which may have been too rich just for plastering. The benefit of this plaster layer is that it acts as a quite effective vapor barrier, which, in combination with the roof membrane (Bituthene® 3000), keeps a comfortable

humidity level not commonly found in most stove-heated log or stackwall homes. In truth, the humidity level is a little too generous at times, and I mop up condensation on the windowsills frequently in the morning. But I look at this as a labor of love; it keeps the sills from rotting and keeps them fresh-looking as well.

The rough plaster was smoothed out with a stucco paint mix made from four parts plaster-board taping compound and one part inexpensive white latex paint. This coating gives the walls the appearance of the lovely adobe buildings in the desert Southwest. It creates a pleasing contrast with the posts, rafters, girders, and exposed plank ceiling, and greatly brightens the interior. To add visual interest, we left the colored bottle ends exposed in these white walls.

The exterior of the house (unplastered cord-wood walls) hasn't caused us any problems. Log shrinkage was minimal (we used white cedar exclusively) and there were very few hairline cracks in the mortar joints, firmly pointed smooth by my wife.

On the roof, the thatch is getting richer with each year, and I still wonder how onion plants found their way up there, in place of the wildflower seed. I guess it gets too dry up there during the hot summer. I don't water the roof. We built the roof as outlined in Rob's book on cordwood-masonry house building, only substituting 6″ of beadboard (two 3″ layers) for the more expensive Styrofoam®, and we've had practically no trouble in the seven years we've lived under it.

The minor exception occurred during the second winter, when a few drops showed up near the outside of the ceiling. The cause turned out to be a tiny air bubble in the overlap of the membrane. It was easy to locate the spot, press down the area of Bituthene® concerned, and apply some Bituthene® mastic. So, extreme caution is advised when laying the membrane. It's probably best to choose a cloudy day to lay the membrane and to give the overlap extra attention. We didn't use the recommended primer with the membrane, but it still seemed to stick down to the wooden roof deck really well.

Our floor is covered with local flagstone, another tedious job best forgotten. We sealed it with an acrylic sealer. I wish now I'd used 2″ of insulation under the slab instead of 1″. I think that would

Illus. 16–6. Siegfried Blum's earth-sheltered cordwood home, in Priceville, Ontario, Canada (Siegfried Blum photo)

Illus. 16–7. The Blums at home

keep the cold stone floor somewhat warmer in winter, although this poses no great problem for us. It's just something that could be improved upon with very little extra expense. In the summer, the floor puts its best foot forward. The delicious coolness to walk on, together with the 'air-conditioned' feeling in the home, make the hot summer months a pleasure to experience.

By now it should be apparent how we'll answer the question at the outset of this story. Yes, we'd do it again, but we're glad we don't have to. It was, at times, a gruelling effort. However, it was a revelation of one's physical and mental makeup. We're happy now to putter around the gardens and feel content with our achievement. The realization that we did all this with our own hands is extremely gratifying!"

BUILDING "EARTHWARD"

by John & Edith Rylander

See the color section for photos of this remarkable home.

"From 1973 to 1987, we lived in a house we built on five acres near a lake in central Minnesota. Our three children eventually moved out on their own, and by 1985 we were alone in a 2000-sq.-ft. home, more room than Edith (poet, novelist, columnist)

and John (an English teacher soon to take early retirement) wanted to deal with.

Our planning, devising, and designing led us across the road, to a wild forty acres we owned. Planning was influenced by our environmental instincts and knowledge, and the design finally coalesced into a small earth-sheltered, post-and-beam, cordwood home. The wing-shaped floor plan has but one bedroom, a galley kitchen, dining area, and living room. A small bathroom and walk-in closet are hidden behind the kitchen wall. We called it 'Earthward,' from Frost's sonnet 'To Earthward.'

As the floor plan shows (Illus. 16-8), 'Earthward' is long and narrow, almost banana-shaped. The end angles are about 15° to the southeast and southwest. The result is that we get not only mid-day sun, but morning and afternoon sun, too.

Digging In

There's a utility room in the rear with ample storage shelving, and access to the garden by a penetrational entrance through the earth berm. A front entryway porch (Illus. 16-9) is screened in summer, enclosed in winter. Altogether, we have 875 sq. ft. of interior space, and figure the home cost about $20/sq. ft., exclusive of our labor.

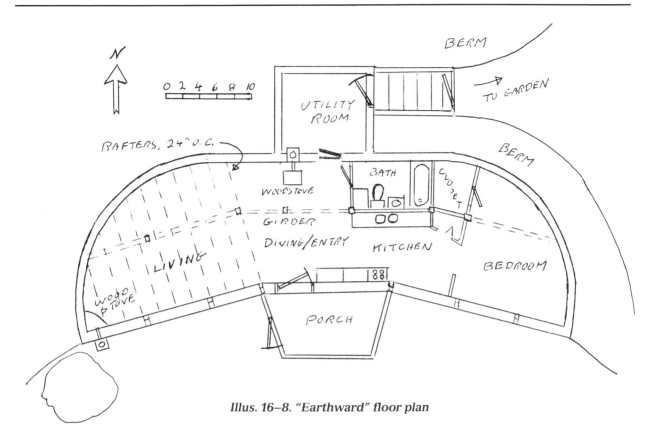

Illus. 16–8. "Earthward" floor plan

Illus. 16–9. "Earthward" (John Rylander photo)

The excavation was made in the autumn of 1986 and we poured the footings for the 55' × 15' structure on June 1, 1987. When our house across the road was suddenly sold, we moved into the weather-tight but very unfinished 'Earthward' that same October.

After six years, we find the advantages of earth-sheltered living fully meet our rosiest anticipations. The southern orientation allows lots of sunlight and solar gain. The roof overhang and Venetian blinds keep out the hottest summer rays. During the record-breaking heat of 1988, we kept interior temperatures between 73° and 80°F, using only a ceiling fan and a box fan.

Electric baseboard heat is available, but it hasn't been turned on except during vacations. The house won't freeze without backup, but we feared that houseplants might succumb during our absence. Our heat comes from us, appliances and lights, and a wood stove. Our firewood is dead or downed wood from our property, yet there's still plenty of tangles and stumps for wildlife shelter. We've never exceeded 2½ cords of wood per year, and seldom need to stoke the fire at night to maintain our 68°–72°F comfort range. Humidity is a comfortable 30–40%. Except for the most humid summer days, we've had almost no condensation problems, and these minor ones were easily cured with a small fan to stir the air around.

Materials

We built 'Earthward' from materials we could provide and work with. The south-facing wall, for example, is cordwood masonry. We'd never seen a cordwood wall or talked to anyone who'd built one, so we practised our techniques on the back wall of a garage, where mistakes wouldn't be too serious. We saw possibilities for both solidity and esthetic satisfaction, possibilities which became realities. The cordwood walls in the home are 20" thick and composed of aspen, oak, red elm, birch, and basswood logs which John had cut from our woodlot.

The large unsplit log ends shrank quite a bit, and we caulked, inside and out, that first year. We used wide mortar joints (painted off-white) with the cordwood. The inside cordwood ends were planed, sanded, and varnished, which brings out the color and grain. Exterior ends were soaked in a wood preservative and old motor oil, so they're dark and contrast well with the paint.

Rafters and beams came mainly from a farm several miles away. The farmer took two crops of hay from our land for fifteen trees from his, a good deal for all. Barter circumvents the need for cash.

For structural and ornamental use, we used a lot of fieldstone, inside and out. Almost anyone can learn to mix 'mud,' handle a trowel, and lay stone in satisfying and structurally strong ways.

Recycled materials (lumber, doors, windows, plumbing, etc.) are much less expensive than new commercial stuff, and they come in greater varieties. We used plenty, including a quantity of neatly stacked lumber which cost about 10% of new material. In short, the use of cordwood masonry, indigenous materials, and recycled goods all helped save us trips to the bank.

Some Relevant Specs

- **Footing:** 3' wide and 4' deep on the front, with 1" of foam insulation on the outside. 2' wide and 3' deep on the back wall.
- **Floor:** 5" concrete slab, with bathroom and kitchen drain lines embedded in the floor.
- **Back wall and utility-room wall:** 12" block with conventional mortar bond, every fourth block cavity slushed with concrete and rebar. Asphalt coating applied to exterior. One-inch foam insulation top to bottom, 2" down to the 4' frost level. Four-inch flexible drainpipe around entire back wall at bottom, middle, and 6" below top.
- **Posts and beams:** 4 × 8 red oak rafters, 24" o.c., ranging up to 16'6" long, and supported by the front and back walls, and an 8 × 8 center beam on 8 × 8 posts.
- **Roof:** 2"-thick white oak rough planking is nailed to the rafters. 2" of rigid-foam insulation (R-11) above the planking. Two layers of cross-laminated Bituthene® come next, and ½" fiberboard protects the Bituthene®. Drainage layer is 2" of gravel covered with a filtration mat. Roof is topped with 9" of black dirt. (The Bituthene® hasn't leaked, but we'd recommend placing it over the roof deck, below the rigid foam. A few mice have found their way into the insulation.)

- **Front wall:** 20″ cordwood infill between posts on 7′ centers, with about 7″ of cellulose insulation laid in between the inner and outer mortar joints. No split wood. 14 different species of wood were used in the 'library wall,' a design feature of the home.
- **Ceiling:** 3½″ fiberglass insulation was stapled between the rafters. Molding strips of ¾″ oak were nailed to the rafters, up 2″ from the bottom. Oak boards were cut to length and placed loose on the molding strips. Estimated insulation value of ceiling and roof is R-30.

Before You Build

Building your own home should start with self-appraisal. A few questions now may save anguish later. Here's what you need to know before you start:

- Can you motivate yourself? Building a home takes a long, steady effort. Will you keep at it, good weather or bad, 'in the mood' or not?
- Have you the physical stamina to see the job through? Weight-lifter strength isn't necessary, but some strength, balance, and general good health is.
- Can you get professional help when needed? We hired pros for some of the block laying, plumbing, wiring, and carpet installation, doing the rest ourselves. We wanted to finish in one summer and were willing to trade money for time.
- Can you organize other people? We hired two recent high-school graduates for much of the summer, and had to plan each day's activities, making sure the right tools and materials were on hand . . . and do our own work at the same time.
- Are your finances adequate? Be sure you'll wind up with a building fit to live in, if not entirely finished, *before you begin*. Discuss your cost estimates with people who have experience in building and building costs. Don't start unless you're sure that you have at least the bare minimum. Lenders are skeptical about owner-builders using unconventional techniques.
- Have you checked local building codes and regulations? Which permits are needed? Which inspections are required, and when? Codes are

written for the standard, not the unconventional. Better to cross every 'T' and dot every 'I' than be forced into expensive changes.
- Can you tolerate frustration? Things don't always go smoothly. A job you thought would take five hours may take five days. Tools break, materials run out or aren't available, or the weather doesn't cooperate. Have a Plan B, even a Plan C. Ulcers will greatly reduce your satisfaction in building your own home.
- If you have a domestic partner, are you sure he/she really wants to get into this? Not every pair that lives together can work together. Mothers-in-law sometimes regard house-building as a form of spouse abuse. House-building isn't a task with which all women will be happy, nor will all men.
- If you're working with a partner who has fewer building skills than you do, can you transmit instruction and criticism in a way which doesn't leave your partner loathing both you and the project? Can you *take* instruction and criticism without too much ego-bruising?
- Can you put up with each other's work habits? Domestic tensions don't improve with the addition of hammers and saws.

Sage Advice

In 1973 we built our first home, starting in May and moving into a very unfinished house in July. At about the same time, folks we knew were building a house about the same size as ours. During construction, husband and wife and two small children lived in a small drafty cabin without indoor plumbing. Five years later they were still in that cabin, while they worked to finish every small detail of their new home, *before* moving in.

Our willingness to move into an unfinished home was to some extent shaped by Ralph, who'd once helped John retrofit a bathroom in an old out-of-square farmhouse where we were living. The last sheet of panelling caused major problems. John and Ralph would trim it, bring it into the bathroom, mark the sheet, take it out, and trim it again. After three tries it was still not exactly right.

Ralph eyeballed the situation and uttered a memorable word of advice. 'John, nail her down. We ain't buildin' no piano!'

If you're building a piano, or a space shuttle, then everything must be perfect. But for amateur builders, doing your best, but settling for the solid and safe, if slightly imperfect, is a valuable lesson. We needed to move into 'Earthward,' and did.

Joining the Underground

We made a conscious choice to go underground. We didn't want the most important investment of our lifetime to encourage clear-cutting of ancient forest, global warming, the production of toxic wastes, or mindless assembly-line labor, anymore than it had to. Building an energy-efficient, low-cost home with mostly local materials, and as much as possible with our own hands, is our statement of how we think human beings ought to live.

Underground house, for many, conjures up images of Grandma's fruit cellar or Tom Sawyer's Cave. It's gratifying to see visitors come in, blink, and say in startled tones, 'Why this, this is . . . nice!' In its comfort and beauty, 'Earthward' argues that ecologically sound housing need not mean shivering in the dark."

17

Case Studies in the Southern U.S.

ROGER DANLEY & BECKY GILLETTE

These two first case studies describe Log End Cave–type houses built in the Deep South. Roger starts the story:

"We'd just finished college in 1981, gotten jobs, and dreamed of sailing around the world. Having no desire for a long-term mortgage, we bought land in rural hilly southern Mississippi, built a pole-barn cabin, and moved in while we decided which house best suited our site.

The day we discovered Rob Roy's book on underground housing was the day our course was set. An earth-sheltered home seemed perfect for us and good for the hilly site. And, by building ourselves, we could escape that dreaded mortgage.

We look back on the years of construction with pleasure. We met many of our friends as a result of our alternative-housing choice. They, too, built their own homes.

Building parties were the order of the day. I'll never forget the roofing party. We must have had 20 people installing those oak planks with 5-pound mini-sledges and 20d cement-coated nails. Our only shortage was crowbars to remove the bent nails. At the end of the day I swept up 20 pounds of nails from the floor.

But the bulk of the work was done by Becky and me (Illus. 17-1). We used very few contractors, mostly for the digging and cover-up. And did we learn some things! For one, a certain brand of planer has a tough time with 5″ × 10″ southern yellow pine and an even tougher time with 2″ × 6″ red oak. Hour after hour was spent feeding that machine.

After months of collecting unwanted cedars from neighbors' lawns, we decided against cordwood masonry. Too many of the cedars we cut had deteriorated sapwood and some heartwood damage. I guess everyone says this about their own region, but the South has a harsh environment. Western red cedar won't last five years outside, even off the ground. In Canada, I've seen 500-year-old logs of the stuff still in good shape. Instead of cordwood, we ended up with cypress shiplap siding milled on the planer, with a 2 × 6 stud wall for extra insulation.

We got a deal on a large sheet of EPDM, a synthetic-rubber membrane, guaranteed for 40 years in an exposed industrial setting. Our only problem has been fire ants, who'll eat a hole right through a 120-mil (⅛″) sheet. So we have to be vigilant in treating any mounds we see on the roof. I don't think it would have made any difference which membrane we'd have used, they would have eaten through it. We mow about twice a year so we can see any mounds that appear. We have a beautiful spring-to-fall wildflower meadow on top.

The first winter after we closed in the house was very severe, − 2°F for days, and below freezing for a month. That may not sound bad, but with the humidity we have . . .

We had a box stove set up in the house shell so we could stay warm, and we started sleeping on an old hide-a-bed in the corner. That was a mistake. Once you move in, work becomes much more difficult. As a result, it took us several extra years to finish.

Illus. 17–1. An early look at Becky and Roger's underground house in the "Deep South" (Becky Gillette photo)

But one day we were finally finished. Cedar planking in the dining room and cedar cabinets in the kitchen. Yellow poplar in the living room and kids' rooms, pine in the pantry, cedar in the mudroom. Blackjack oak and cherry molding in the bathroom. Ceramic tile throughout. A very warm and earthy place, all-electric with an average bill of $45 a month.

With a west-facing exposure, we were forced to put a porch in front to shade the face of the house. One sunny afternoon, while sitting with friends on the porch, a man dropped by with an aerial photograph of our place. He had mistakenly centered the cabin, as the house was nearly invisible from the air! How's that for blending into the landscape? After 12 years, our cattle pasture had turned into a beautiful forest. The most glaring object in the photo was our 22' sailboat, sitting unmoved these last five years. Oh, yes, the dream. Sitting there that day, we realized we still had it.

So here we are a year later, moved 90 miles to the coast, where we built a house on 8 × 8 piers. We have a long-term lease on the earth-house, to people who really appreciate and take care of the place. Someday it will be our retirement home. In the meantime, the boat is in the water, we sail regularly, and we're looking for 'the' boat.

In conclusion, I'd like to thank Rob Roy for all he did for us, giving us the confidence to do what we did. I don't think we would have our favorable financial situation had we bought into a mortgage, so the hard work was worth it. As it is, we'll use the income from the earth house to further our sailing dream."

Becky adds some hard facts:

"In a nutshell, here's what we've learned about earth homes in the South:

- Since the ground temperature averages 70° to 72°F, the earth home didn't stay cool or dry enough without air conditioning. We live in a hot, humid environment, and things like shoes in the closet tended to mold, so we added a small window air conditioner in the back room. It didn't use much electricity, but made the house comfortable. Our total electric bill (no gas appliances) averaged less than $50 a month year-round, ranging from $90 in the summer to $30 in the winter. If you're building *only* for energy savings, an earth home may not be the right choice. Our stilt house on the coast has 2 × 6 insulated walls and low-E glass and has summer electric bills as low as those for the earth home. But energy efficiency isn't the only consideration. Tornadoes have struck within a couple of miles of us quite frequently, including a huge one that wiped out an entire community. Then there was the aftermath of Hurricane Andrew. In the middle of the night, after high winds had knocked the power out, my brother-in-law, James, staying in the pole cabin, came banging on the door. When the cabin started swaying, James headed for safety like a rabbit into a burrow.
- Roger mentioned the fire ants. They caused leaks in our EPDM membrane, and it's a pain to fix leaks on a sod roof. A leak doesn't always drip exactly where the hole is, so leaks are hard to find. You can dig up a large area to find a small hole. With 60" of rain a year, and as much as 8" in a day, even small holes can be big problems. You can kill all the ants, and new colonies will spring up overnight. I don't know the answer. A nearby home was earth-sheltered with a conventional roof, and it looks funny. Our earth home looks much nicer with the sod roof, especially when the daffodils and coreopsis are in bloom. The house blends into the environment so well that people drive up and ask, 'Where's the house?'
- With our warm ground temperatures, we can have windows that open. In the Southern earth home we don't have the problem of condensation on the walls.

- While most owner-builders probably don't plan to sell, if you know in advance that you will, you'd be safer with a conventional home. Most banks sell their mortgages to Fannie Mae (Federal National Mortgage Association), which requires three comparable sales as reference. There weren't three comparable earth-home sales in our area, hence it *may* have been difficult to get bank financing. We could have owner-financed, of course, as we didn't have a mortgage.

- When potential buyers came to look at the house, the common response was that they loved how it looked, but were scared about it being different, scared they might not be able to get insurance (we finally bought insurance ourselves just to show it could be done), scared they couldn't re-sell if necessary. Now that the home is being leased, we've had several opportunities to sell it, but we now want to keep it for our retirement when the coast gets too crowded.

- Our total building cost (excluding the value of our labor) was about $50,000. That included 4.2 acres of land, a pole-frame cabin we lived in for two years, several sheds and a shop, a well and septic system. By building as we could afford it, and only taking out small short-term loans, the house was paid for when it was finished."

LINDA & RAY HURST

Linda and Ray's home in Alabama is less than 100 miles away from Becky and Roger's in a virtually identical climate that's not far from the Gulf of Mexico. In a 1986 letter, Linda wrote:

"It was such a pleasant surprise to open the mailbox and find a letter from our mentor. You may be sorry you asked about the house. I, too, could write a book! Bottom line: We love the house. But we've had our pitfalls.

We used your Log End Cave plan, but with dimensions of 35' × 54'. Later, we added a 25' × 15' greenhouse on the south side. We probably shouldn't have faced the house west, but the view was too good to pass up. In the summer months, we put up solar screens to cut out the direct heat from the evening sun. In the winter, the sun helps warm the house, but staying warm is no problem. The *coolest* the house has ever been is

72°F. December, January, and February are the coldest months, with temperatures in the teens not uncommon. When it's nippy, we fire the wood-stove, perhaps 8 or 10 times during the winter.

The heat in the summer is another matter. Actually, Ray and I were fairly comfortable. The house would heat up to 84°F, while outside temperatures were in the high 90s. We have ceiling fans in every room; so as long as the air was moving we were okay. However, our 'city folk' family wouldn't come to see us in the summer because of the heat, so we gave in and installed air conditioning in 1985. I keep the setting on 80°F. Just that little difference, plus the dehumidification, makes it much more pleasant inside. I must admit that even I enjoy the a/c now. The unit runs infrequently, and our monthly power bill goes from $50 to $80. Humidity is a problem in the South. We use a dehumidifier in the winter. But a great many 'conventional' homes here have more moisture problems than we do. We've never had a mildew problem and many of my friends do."

But they did have some minor roof leaks, according to Linda's letter of 1986, and termite problems. They'd used a combination of waterproofing techniques, including black plastic and bentonite clay, but they had to dig up the earth at the edge of the house, patch it and, importantly, fill in with gravel so that the water could find its way to the footing drains. Ray's view was, "I built the thing, I can fix it!"

Like Becky and Roger, Linda reported that she and Ray moved into the house the minute the roof was on, covered temporarily with felt. They lived in "total confusion and disarray for what seemed like years."

They removed the crosstie retaining walls, which had become infested with termites. "This was too close to the house for comfort," said Linda. "We replaced the ties with surface-bonded block walls."

Linda and Ray built their home during the early 1980s, when underground housing had captured the public's imagination. "One drawback," Linda wrote, "is the endless onslaught of the curious. They come from nowhere, people we've never heard of, even people from other states. It was kind of neat at first, but we have long since tired of having our solitude invaded. Every soul who has ever even thought about building one of these comes, picks our brains, and calls for advice, although they seldom follow it."

In 1993, Linda wrote: "Building this place seems like a vague memory. We don't remember the hard parts or the fact that it took years. It was quite an undertaking for two city dwellers with little knowledge of building. I was a secretary and Ray worked on a river barge. But I'm glad we did it. We have great financial freedom now in our retirement as a result of no mortgage and very low upkeep and utility bills. We've also planted a seed in the minds of the boys, who both plan to build their own homes just as soon as they get their degrees and get 'real' jobs.

We used cedar for the cordwood masonry. It's still beautiful inside, but we should have sealed the outside. We have so much rain and moisture in the South that the logs have mildewed and turned dark on the outside. I think we could use bleach and a brush to get it off, and then seal it, but we've got too many irons in the fire to worry about it."

Linda reported further problems with termites ("insidious little beasts"):

"We chose not to treat the ground under and all around us (like most people do here) because we didn't want to surround ourselves with poison. But we had a problem with termites on the roof. Fortunately, we knew what to watch for and got to them before any significant damage was done. However, we chose at that time to remove the dirt, and replaced it with a more conventional roof. We built the walls two courses higher, and then added 2×10s on 24" centers to make the roof even with the blocks. We added insulation, decking, and shingles. So the surrounding earth is 10" below the roof. Once a year, now, Ray goes around the perimeter and sprays the ground and under the eaves. This spraying has worked fine.

I guess it's a trade-off of sorts. I wanted the beauty of the wood in the ceiling, but I wonder if it wouldn't be better to choose something termite-proof for the roof and ceiling. Can't use treated wood: more poison. I think sometimes we wind up harming ourselves more than we harm the termites!

There's another fellow here in town who had an underground home professionally built. I've noticed that he, too, has taken the dirt off his roof. I'm inclined to think he had a similar experience."

AUTHOR'S COMMENTARY

It's enlightening to hear from people like Becky and Roger, and Linda and Ray. It's so easy to forget that climate and local conditions can make such a huge difference in the way a home should be designed and detailed. While the basic construction methods described in this book should hold up throughout the U.S., I admit to having little experience with destructive ants and termites. Once, about ten years ago, a band of carpenter ants made a tiny hole in our Bituthene®. We repaired the hole with a patch and eradicated the ants, and we've had no problem since. The Earthwood roof has the advantage of being freestanding. The earth doesn't berm up onto the roof, as in the Cave-type design, so it's hard for undesirable wildlife to gain access. Maybe a freestanding earth roof with a termite-shield drip edge would be a good idea in the "Deep South." Seek local advice from your county or town building department.

If you live in the southern U.S., I hope that you, too, will gain from the experiences of these courageous people.

OUR UNDERGROUND HOUSE

by Geoff Huggins

"Here in Virginia an underground house faces a mixed climatic situation. Its design must contend with both cold winters and hot, humid summers. Thus, one's motivation for going underground could be justified by hoping to save either heating energy in the winter or cooling energy in the summer. Since we (Louisa Poulin, my construction partner, and I) happen to have plenty of firewood on our land, the scales were tipped towards summer considerations. Design choices thus were often driven by a concern to keep our house cool in hot weather.

We chose to use earth-bermed walls and an earth roof for three reasons. (1) The earth layer provides a buffer that greatly reduces seasonal

Illus. 17–2. Geoff and Louisa's underground home in 1993 (Geoff Huggins photo)

Illus. 17–3. Geoff with his power compactor. He supplied the power. (Louisa Poulin photo)

temperature fluctuations at the roof. A roof exposed to air sees seasonal temperature fluctuations of at least 110°F in our area, while submergence under only 4″ of soil reduces that seasonal fluctuation to about 40°F. (2) Because earth doesn't insulate (it *stores* heat), solar energy absorbed by soil during hot weather can be returned for use during cold weather. (3) Vegetation growing on the roof provides a natural cooling process during hot summer days, due to evapotranspiration. (Plants pull moisture from the soil and evaporate it into the air.)

Our house (Illus. 17-2) is 25′ × 40′ (1000 sq. ft.), inside dimensions. There are no interior walls, and ceilings are 10′ high, which makes the living space seem larger than it is. Illus. 17-3 shows me at work preparing the sand pad prior to the floor pour. For good thermal contact with the earth, the floor is a cement slab. Above-ground walls are constructed of cordwood, while underground walls are dry-stacked concrete block, coated with surface-bonding cement. Liking the warmth of wood, we chose not to have a concrete ceiling, and instead designed 6″ × 12″ rafters (at somewhat less than 4-foot spacing) supported on a 12″ × 12″ beam (Illus. 17-4) down the center of the house. Both beam and rafters were assembled by bolting together 3″ × 12″ rough-sawn oak timbers. Using small components made the weight of any one member manageable enough that two people could lift it for assembly.

Two retaining walls hold back the earth (Illus. 17-5): a short mortared-stone wall at the southwest corner and a longer concrete-block (surface-bonded) wall midway on the east wall. The longer wall is supported by six tiebacks (Illus. 17-6), which were made by attaching one end of steel cables (through small holes in the back of the block) to reinforcing bar inside the wall, and the other end to large stones buried in the backfill.

Large windows in the south and east walls provide abundant natural light to most of the house. A clerestory over the buried northwest corner brings much-needed light to an otherwise dark location. Roof penetrations are held to a minimum by having water and cooking-gas lines and the septic ventilation stack share the clerestory structure. There's also an air-ventilation door in the clerestory to allow warm air in the house to rise

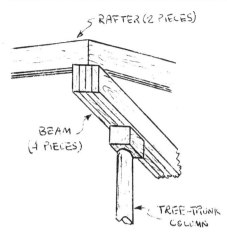

Illus. 17–4. The main girder (beam), as well as all rafters, was made of 3″ × 12″ rough-sawn oak.

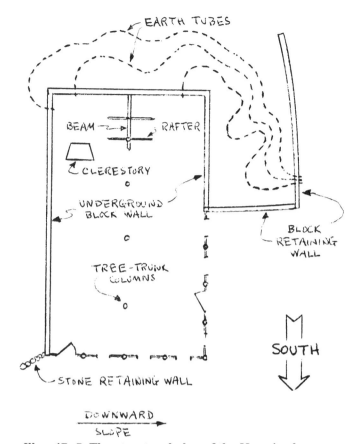

Illus. 17–5. The structural plan of the Huggins home also shows the location of retaining walls and earth tubes.

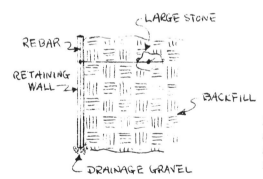

Illus. 17–6. Left: "Tiebacks" made of cable attached to large stones resist earth pressure against the retaining wall, much as the "dead men" do (Illus. 13–2).

and flow outdoors. Fresh inlet air is provided by three earth tubes entering the back walls of the house, down near the floor level (as well as a separate outside air-feed supply for the wood stove).

A major challenge here in Virginia is condensation in the summer. Although we rarely get moisture actually condensing on the lower block walls, humidity throughout the house remains high all summer. On those sultry days when the outdoor temperature is 95°F and the relative humidity is high, when some of that air moves indoors it's cooled to about 75°F as its heat is absorbed into the mass of earth behind the walls. A lot of water vapor is dropped in the process. We must run a

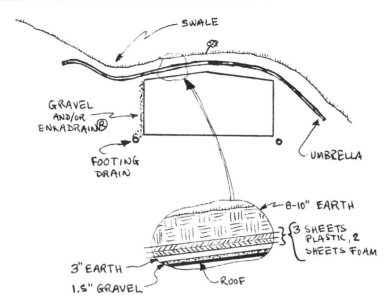

Illus. 17–7. Details of Geoff's insulation/watershed umbrella

dehumidifier 4 to 5 hours each day on hot and humid summer days. Somewhat in compensation, our interior humidity in the winter rarely falls below 65%.

A major feature of our earth-cover design is the insulation/watershed umbrella (Illus. 17-7) we constructed over the house. This concept was developed by John N. Hait, based on research done by the University of Minnesota and described in his book *Passive Annual Heat Storage*. The umbrella is a multilayered treatment of 3 sheets of 6-mil polyethylene sandwiching 2 layers of 2"-thick expanded polystyrene (beadboard) insulation. This 'umbrella' approach appealed to us because it simultaneously tackles 'waterproofing,' insulation, and heat storage.

I use the word 'waterproofing,' although I believe that there's no such thing as a permanently waterproofed underground structure; sooner or later, moisture can be expected to breach the defenses. We implemented a 'water-resistant' approach that consists of a 4-step process (starting with the most important): (1) surface water is encouraged to flow away from the house by constructing appropriate slopes and swales; (2) the umbrella sheds water that seeps into the soil, keeping the earth dry below it; (3) air spaces are provided by gravel layers and plastic fibre mats immediately against the roof and walls, such that any water that may someday somehow penetrate the umbrella will freely fall to the footing drains and be carried away; and (4) the outside of the roof and walls are surfaced with asphalt roof coating and 6-mil polyethylene sheets (primarily as a vapor barrier, not as a bulk-moisture defense).

Note that the roof and walls of the house aren't insulated from the surrounding earth, but are in direct thermal contact with it. It's the house *and its* surrounding earth ball that are insulated by the umbrella from the wide seasonal temperature variations of the air. If the walls are in thermal contact with this adjacent earth ball, heat can readily flow into the earth during the summer (keeping the house cool), get stored in all that mass of dirt, and be released back into the house during the winter, sort of an annual heat exchanger. John Hait points out that it's critical that the earth stay dry. Wet earth, rather than storing heat for later use, will simply conduct it away. A main function of the umbrella is to keep the soil dry (in addition to its insulating role).

A few words on how we did it: The early first step was lots of research. We read every book and periodical on underground construction we could find. There isn't an extensive literature, so informing oneself is not a gargantuan task. We took Rob Roy's hands-on building courses, and I took an applied building-science course. The value of a sound education cannot be overstated. We began as neophytes, boot-strapping ourselves up from ground zero. We'd never built a structure before. We tried a small cabin as proof we could pull it off, and are now living in our second structure: our underground house.

Another phase whose importance is hard to overstate is the design/planning stage. Don't abbreviate this stage. The complexity of the whole process seems overwhelming, especially to those who've never built before. It's a complex procedure of many, many steps. Keep in mind, though, that through careful planning the process can be reduced to a sequence of simple steps; line them up and take the first one.

If you have a building code to contend with, deal with your local building officials with care, consideration, and honesty. They hold the power; don't buck them, but go with the flow. Compliance with the code can be a particular challenge when building an underground house, since your local code-enforcement official may be unfamiliar with the techniques you plan to use. Consult the permit department very early in the process, and find out everything you can about the law. The department will have pamphlets and books to inform you about regulations. By thoroughly planning your underground house beforehand, you can impress your code officials with your sincerity. Have the building well planned, but not to unreasonable detail. Use the expertise of the building officials to help design final details. Get them involved in the process. For example, the plumbing inspector helped me with drain-line details. When he subsequently came to the site to perform the inspection, he hardly looked at the system, because he was already familiar with how it would look. Give your code official credit for being cooperative. An ounce of friendly persuasion in his office can avoid a pound of grief when he appears on site to inspect.

Louisa and I were fortunate to have good jobs in the city, which allowed us to build some savings. We were able to leave those jobs, nearly 10 years ago, move out to the country and build full-time, while residing in our cabin. The savings freed us from the need to borrow construction money. We did all the work ourselves, except digging the hole into the hillside and finishing the concrete slab. (The thought of 14 yards of concrete being dumped on the ground and wanting to be evenly spread out in a couple hours' time was intimidating to me.) We've been building the house for over 9 years now, and living in it for 7 years. It is about 80–90% complete and very livable. Does it seem as though we should be further along, after nearly 10 years? Maybe so. It used to concern us, at times. But then we chanced upon an old Chinese proverb that evaporated this concern, and which we have since adopted: 'Man who finish house, die.'"

Appendix 1: Radon

Several years ago my good friends Peter and Eileen Allen built a wonderful earth-sheltered home. The home features two "double domes," built on 4' earth-bermed knee walls. A 26'-diameter dome is constructed within a 29' dome, the geometries matching exactly. The space between the dome frameworks is filled with sheetrock, a plastic vapor barrier, 12" of fiberglass insulation, and a vented air space. The vapor barrier is quite tight, as are the concrete knee wall and floor, except as noted below. Heat is delivered to the concrete floor by a wood-fired in-slab conduction-and-convection system, carefully laid out by Eileen's father, dome designer George Barber. It was decided to build the home over the well, so that there would never be any concern about frozen pipes during severe winters.

A couple of years after building the home, Peter's brother, a physician, made a gift of a radon-testing device to the young couple, concerned, perhaps, about some of the construction techniques used. It may have been the best present they ever received. When the test results came back, they learned that the level of radon in the winter in the home was about 40 picocuries per litre (pCi/l), or ten times the "action level" recommended by the U.S. Environmental Protection Agency. At 4 pCi/l, the EPA recommends that remedial work be done to lower the levels of radon.

Radon's greatest health threat is increased risk of lung cancer. For example, breathing 20 picocuries per litre is considered to increase the risk of lung cancer about as much as smoking two packs of cigarettes a day. Pete and Eileen, organic gardeners and clean-air freaks, didn't much like the idea of having 30 times more risk of lung cancer than a nonsmoker living in a low-radon house.[11]

About the time that Peter and Eileen were battling radon, Jaki and I were thinking of moving our oldest son, Rohan, to a bedroom in the fully earth-sheltered part of Earthwood, down on the first floor, in order to make room for little brother Darin in Ro's upstairs room. We were particularly concerned because we'd included into the walls of the home earth tubes for summer cooling. We wanted to be sure that we weren't going to subject a 9-year-old to the equivalent of smoking a pack of cigarettes every night while he slept. We sent for a test kit, and were relieved to receive an analysis of just 0.4 pCi/l, or 1/10 of the EPA's action-level figure.

Lucky for us, but what about our friends? They tested again in the summer and found that radon levels were only 1 or 2 pCi/l, well within safety limits, but in winter, levels were high again. Remember that their dug well was right in the house, and quite close to the location of the wood stove which powered the in-slab heating system. Peter observed that cobwebs near the top of the well were rising up substantially whenever the wood was burning. The wood stove was creating a negative pressure in the soil gases as the stove tried to find air to replace the air that was being exhausted by the chimney. Given the seasonal nature of the radon problem, it was obvious that the wood stove pulling soil gases from the well was the culprit.

The last upper ten feet of the well was a 30"-diameter metal culvert. Peter was able to seal off the well by pouring concrete in the bottom of the culvert. In addition, he sealed up other direct exposures to soil gases, such as a house-drain cleanout. And, to help decrease the negative pressure while the stove was in use, he opened a second direct-air inlet to the outside. The efforts paid off, and the home's winter radon levels are now at or just under the EPA's recommended action level. The Allens figure that the average level for the year is about 2 pCi/l. Eileen asked me to emphasize that, after construction, any airtight house should be tested for radon gas.

Is radon a serious problem? The EPA estimates that eight million American homes may contain unsafe levels of radon. The National Cancer Institute considers radon exposure to be the greatest cause of lung cancer after smoking, accounting for 30,000 deaths a year in the U.S.[12]

What are the ramifications for earth-sheltered housing? Michael Lafavore, writing in his book *Radon: The Invisible Threat*, says:

Earth-sheltered and underground homes have a larger surface area in contact with the soil, which means more places for radon to enter. "We've yet to see an underground house that doesn't have some radon in it," says B.V. Alvarez, president of Airchek, a radon-testing company.[13]

You should find out if the soils on the site where you intend to build hold a lot of radon gas. Generalized maps of radon "areas" are only somewhat helpful, however. Sometimes, one building lot may have high levels of radon and another one 100 yards away may have no radon at all. The only way to be sure is to conduct a soil test. One company which supplies a site-testing kit is Airchek. They'll mail you the kit and provide an analysis after you mail the charcoal packet back to them. Set up the little cardboard tent cubicle they supply on the subsoil itself, so if you're planning on a deep-hole test for septic-system purposes, this would be a good time to check for radon.

Your state health department is a good source for getting names of reputable testing agencies in your area. Also, the EPA has two useful (and free) pamphlets on radon, *A Citizen's Guide to Radon* and *Radon Reduction in New Construction*.

What if the site is "hot"? There are techniques that allow you to build on a site where there's radon in the soil. You want to be sure that the house is completely sealed against infiltration of soil gases. The Bituthene® membrane will help a great deal, as will a 4″ reinforced-concrete slab with 10-mil black polyethylene beneath it (in addition to any rigid foam). Avoid expansion joints between the floor and footings, as this will provide a place for infiltration of contaminated soil gases. Strictly speaking, the 40′ × 40′ Log End Cave plan is the largest home which can be built without expansion joints in the concrete. If you must include expansion jointing, be sure to tool and caulk the joint thoroughly.

If you're planning on wood heat, be sure to provide a positive source of outside combustion air to the stove. This pipe should be a 6″ solid-walled pipe, fully glued, and sealed as it enters the home.

There's a chapter on radon in *Wood Frame House Construction* (Sterling Publishing Co., Inc., 1992). The chapter includes several pages of detail on both prevention in new construction and mitigation in existing buildings. This is the best and most up-to-date information I've seen on the subject.

Finally, there's a product called Enkavent® (a thicker version of AKZO's Enkadrain® drainage product discussed in chapter 10), which is installed under the floor pour. Enkavent® provides an air space that intercepts radon before it seeps into the home. If concentrations of the gas are discovered after the house is completed, pipes can be added to route the gas outdoors.

In conclusion, know the enemy. Find out if you're in a high-risk area. The local or state health department or cooperative-extension agent may be able to help. Conduct a site test, particularly if you're in an area with granite, shale, or phosphates near the surface. If necessary, incorporate radon-fighting details at the design stage. It's hard to retrofit an underground home for radon protection. Build with informed confidence, and don't allow smoking in the home. Radon actually grabs onto smoke particles in the air, exacerbating the problem.

Appendix 2: Sources

ORGANIZATIONS

Underground Space Center, 790 Civil and Engineering Building, 500 Pillsbury Dr. SE, Minneapolis, MN 55455. Director: Raymond Sterling. Associate Director: John Carmody. Sterling and Carmody have collaborated on a number of books on earth-sheltered housing, most of which are out of print. The center does photocopy some of these books.

American Underground-Space Association, 511 11th Avenue South, Box 320, Minneapolis, MN 55415. "The AUA is an organization of professionals involved in every aspect of the underground construction industry. More than 40 professional disciplines are represented in its membership, including engineers, contractors, suppliers, researchers, and architects." While the AUA has been dealing mostly with commercial underground space and tunnelling during the past ten years, it reports a recent increase of interest in underground homes.

British Earth Sheltering Association, Caer Llan Berm House, Lydart, Monmouth, Gwent, Wales, U.K. NP5 4JJ. Director: Peter Carpenter. "BESA is a non-profit organization aiming to encourage the design and construction of earth-sheltered buildings in the U.K." It publishes at least three journals a year. Membership dues include the journals.

AUTHORS, ARCHITECTS

Malcolm Wells, author, architect. Underground Art Gallery, 673 Satucket Road, Brewster, MA 02631. Ask Wells for his current book list. He's also available for architectural services and consultations.

Don Metz, Metz & Thornton, Architects, P.O. Box 52, Lyme, NH 03768. Architect Metz has designed a number of earth-sheltered homes.

INSTRUCTION

Earthwood Building School, 366 Murtagh Hill Road, West Chazy, NY 12992. Rob Roy, Director. Earthwood has offered workshops in earth-sheltered housing and cordwood-masonry construction since 1980, and the school acts as a books-and-plans clearinghouse in these areas.

PRODUCTS

Surface-Bonding Cements

W. R. Bonsal Co., P.O. Box 38, Lilesville, NC 28091. *Surewall® Surface Bonding Cement.*
Conproco, 24 Industrial Park Dr., Hooksett, NH 03106. *Foundation Coat®.*
Quikrete, 1790 Century Cir., Atlanta, GA 30345. *Quikwall® Cement.*
Stone Mountain Manufacturing Co., Lafayette Ctr. Suite 304, P.O. Box 7320, Chesapeake Blvd., Norfolk, VA 23509. *Fiberbond®.*

Waterproofing Membrane

W. R. Grace and Co., 62 Whittemore Ave., Cambridge, MA 02140. *Bituthene®.*

Drainage, Radon-Venting, & Radon-Detection Materials

Airchek, Box 2000, Arden, NC 28704
AKZO Industrial Systems Co., P.O. Box 7249, Asheville, NC 28802. *Enkadrain®, Enkavent®.*
Dow Chemical Co., 2020 Willard H. Dow Ctr., Midland, MI 48674. *Styrofoam® Therma-Dry®.*

Bonding Agent

Thoro System Products, 7899 NW 38th St., Miami, FL 33166. *Acryl-60®.*

Wire Mold

Wiremold Co., 60 Woodlawn St., West Hartford, CT 06110

Footnotes

[1]Underground Space Center, Univ. of Minnesota, Carmody, John, and Sterling, Raymond, *Earth Sheltered Housing Design:* Second Edition, Van Nostrand Reinhold Co., 1985, pp. 11–12.

[2]Wells, Malcolm, *How to Build an Underground House,* Brewster, MA, 1991, p. 3.

[3]Underground Space Center, Univ. of Minnesota, *Earth Sheltered Homes: Plans and Designs,* Van Nostrand Reinhold Co., 1981, p. 96.

[4]Oehler, Mike, *The $50 and Up Underground House Book,* Mole Publishing Company, Bonners Ferry, ID, 1992, p. 9.

[5]Underground Space Center, *Earth Sheltered Housing Design,* p. 238.

[6]Campbell, Stu, *The Underground House Book,* Garden Way Publishing, Charlotte, VT, 1980, p. 21.

[7]Wells, Malcolm, *Notes from the Energy Underground,* Van Nostrand Reinhold Co., 1980, p. 131.

[8]Wells, *How to Build an Underground House,* p. 51.

[9]Ibid., p. 52.

[10]Dempewolff, Richard F., "Your Next House Could Have a Grass Roof," *Popular Mechanics,* March 1977, p. 144.

[11]Lafavore, Michael, *Radon: The Invisible Threat,* Rodale Press, Emmaus, PA, 1987, p. 87.

[12]Ibid., p. 60.

[13]Ibid., p. 58.

Bibliography

Most of the books listed below are available from the Earthwood Building School.

Alth, Charlotte and Max. *Wells and Septic Systems.* Blue Ridge Summit, PA: TAB Books, 1992.

Alth, Max. Do-It-Yourself Plumbing. New York: Sterling Publishing Co., Inc., 1987.

Armpriester, K. E. *Do Your Own Wiring.* New York: Sterling Publishing Co., Inc., 1991.

Hait, John N. *Passive Annual Heat Storage.* Missoula, MT: Rocky Mountain Research Center, 1983.

Heldmann, Carl. *Be Your Own House Contractor.* Pownal, VT: Garden Way Publishing, 1986.

McClintock, Mike. *Alternative Housebuilding.* New York: Sterling Publishing Co., Inc., 1984.

McRaven Charles. *Building with Stone.* Pownal, VT: Garden Way Publishing, 1989.

Nearing, Helen and Scott. *Living the Good Life.* New York: Schocken Books, 1987.

Oehler, Mike. *The $50 and Up Underground House Book.* Bonners Ferry, ID: Mole Publishing Co., 1992.

Roy, Rob. *Complete Book of Cordwood Masonry Housebuilding: The Earthwood Method.* New York: Sterling Publishing Co., Inc., 1992.

Sherwood, Gerald E., and Robert C. Stroh. *Wood Frame House Construction: A Do-It-Yourself Guide.* New York: Sterling Publishing Co., Inc., 1992.

Sobon, Jack. *Build a Classic Timber-Framed House.* Pownal, VT: Garden Way Publishing, 1994.

U.S. Dept. of Health & Human Services. *A Citizen's Guide to Radon.* Washington, DC: U.S. G.P.O., 1992.

U.S. Environmental Protection Agency. *Radon Reduction in New Construction.* Washington, DC: U.S. G.P.O.

Wells, Malcolm. *An Architect's Sketchbook of Underground Buildings.* Brewster, MA: self-published, 1991.

———. *How to Build an Underground House.* Brewster, MA: self-published, 1991.

OUT-OF PRINT TITLES

Try to find these books in your public library or in a used-book store. Some Underground Space Center books are available from the Center in photocopied form.

Campbell, Stu. *The Underground House Book.* Charlotte, VT: Garden Way Publishing, 1980.

Lafavore, Michael. *Radon: The Invisible Threat.* Emmaus, PA: Rodale Press, 1987.

Underground Space Center, University of Minnesota. *Earth Sheltered Homes: Plans and Designs.* New York: Van Nostrand Reinhold Co., 1981.

———. *Earth Sheltered Housing Design.* Second Edition. New York: Van Nostrand Reinhold Co., 1985.

———. *Earth Sheltered Residential Design Manual.* New York: Van Nostrand Reinhold Co., 1982.

Wells, Malcolm. *Notes from the Energy Underground.* New York: Van Nostrand Reinhold Co., 1980.

Metric Equivalents

LENGTH

1 centimeter = 0.3937 inch	1 inch = 2.54 centimeters
1 meter = 39.370 inches	1 inch = 0.0254 meter
1 meter = 3.2808 feet	1 foot = 0.3048 meter
1 meter = 1.0936 yards	1 yard = 0.9144 meter
1 kilometer = 0.6214 mile	1 mile = 1.6093 kilometers

AREA

1 square centimeter = 0.155 square inch	1 square inch = 6.45 sq cm
1 square meter = 10.764 square feet	1 square foot = 0.092 sq meter
1 square meter = 1.196 square yards	1 square yard = 0.836 sq meter
1 square kilometer = 0.386 square mile	1 square mile = 2.590 sq km
1 hectare = 2.471 acres	1 acre = 0.405 hectare

VOLUME

1 cubic centimeter = 0.061 cubic inch	1 cubic inch = 16.387 cu cm
1 cubic meter = 35.314 cubic feet	1 cubic foot = 0.028 cu meter
1 cubic meter = 1.308 cubic yards	1 cubic yard = 0.764 cu meter

CAPACITY

1 liter = 0.0338 U. S. fluid ounce	1 U. S. fluid ounce = 29.586 ml
1 liter = 1.0567 U. S. liquid quarts	1 U. S. liquid quart = 0.946 liter
1 liter = 0.9081 U. S. dry quart	1 U. S. dry quart = 1.111 liters
1 liter = 0.2642 U. S. gallon	1 U. S. gallon = 3.785 liters

MASS OR WEIGHT

1 kilogram = 2.2046 avoirdupois pounds	1 pound = 0.4536 kg
1 metric ton = 2,204.6 avoirdupois pounds	1 pound = 0.000454 metric ton
1 metric ton = 1.1023 short tons	1 short ton = 0.97 metric ton
1 metric ton = 0.9842 long ton	1 long ton = 1.10 metric tons

TEMPERATURE

To convert from Celsius to Fahrenheit, and vice versa, the following formulas are used: $°F = 9/5°C + 32$, and $°C = 5/9 (°F - 32)$. For example, to convert 20°C to Fahrenheit, multiply 20 by 9 (equals 180), divide 180 by 5 (equals 36), and add 32 (equals 68°F). To convert 68°F to Celsius, subtract 32 from 68 (equals 36), multiply 36 by 5 (equals 180), and divide 180 by 9 (equals 20°C).

Index